# Stepwise Culture of
# Human Adult Stem Cells

# RIVER PUBLISHERS SERIES IN BIOTECHNOLOGY AND MEDICAL RESEARCH

*Editors-in-Chief:*

**PAOLO DI NARDO**
*University of Rome Tor Vergata,*
*Italy*

**PRANELA RAMESHWAR**
*Rutgers University,*
*USA*

**ALAIN VERTES**
*London Business School,*
*UK and NxR Biotechnologies,*
*Switzerland*

Aiming primarily at providing detailed snapshots of critical issues in biotechnology and medicine that are reaching a tipping point in financial investment or industrial deployment, the scope of the series encompasses various specialty areas including pharmaceutical sciences and healthcare, industrial biotechnology, and biomaterials. Areas of primary interest comprise immunology, virology, microbiology, molecular biology, stem cells, hematopoiesis, oncology, regenerative medicine, biologics, polymer science, formulation and drug delivery, renewable chemicals, manufacturing, and biorefineries.

Each volume presents comprehensive review and opinion articles covering all fundamental aspect of the focus topic. The editors/authors of each volume are experts in their respective fields and publications are peer-reviewed.

For a list of other books in this series, visit www.riverpublishers.com

# Stepwise Culture of Human Adult Stem Cells

### Editor

## Pranela Rameshwar

Rutgers University, USA

Routledge
Taylor & Francis Group
NEW YORK AND LONDON

**Published 2024 by River Publishers**
River Publishers
Alsbjergvej 10, 9260 Gistrup, Denmark
www.riverpublishers.com

**Distributed exclusively by Routledge**
605 Third Avenue, New York, NY 10017, USA
4 Park Square, Milton Park, Abingdon, Oxon OX14 4RN

*Stepwise Culture of Human Adult Stem Cells / Pranela Rameshwar.*

Routledge is an imprint of the Taylor & Francis Group, an informa business

ISBN 978-87-7022-854-1 (hardback)
ISBN 978-87-7004-097-6 (paperback)
ISBN 978-10-4001-077-8 (online)
ISBN 978-1-003-46687-1 (ebook master)

While every effort is made to provide dependable information, the publisher, authors, and editors cannot be held responsible for any errors or omissions.

# Contents

**5 Isolating Compact Bone-derived Mesenchymal Stem Cells from Rodent and Rabbit Femurs**     **37**

*Najerie McMillian, Joseph S. Fernandez-Moure, and Taylor Hudson*

**6 Isolating Dental Pulp Stem Cells**     **45**

*Ioanna Tsolaki, Darling Rojas, Adam Eljarrah, and Pranela Rameshwar*

**13   Culturing Human Bone Marrow Stromal Cells**        **101**
*Yannick Kenfack, Lauren S. Sherman, Bobak Shadpoor,*
*Andrew Petryna, Sami Souyah, Ella Einstein, Stephanie Perricho,*
*and Pranela Rameshwar*

**14   Isolating Mononuclear Cells by Ficoll-Hypaque**
**Density Gradient**        **105**
*Andrew Petryna, Lauren S. Sherman, Anushka Sarkar,*
*Bobak Shadpoor, Yannick Kenfack, and Pranela Rameshwar*

**15   Phenotypic and Multipotent Characterization of**
**Bone-Marrow-derived Mesenchymal Stem Cells**        **111**
*Anuska Sarkar, Andrew Petryna, Bobak Shadpoor,*
*Lauren S. Sherman, Pranela Rameshwar*

# Preface

After decades of research and clinical trial on fetal, embryonic, and adult stem cells, the larger scientific and clinical communities seem to favor the development of translational studies using adult stem cells. One of the impediments for stem cells and their products to progress to the clinic is partly due to a lack of consensus on standardized methods. The scientific literature has determined that stem cell applications also include the secretome and other products derived from various stem cells that can influence the tissue niche. An example of secretome is microvesicles that can actively transfer information among cells. Thus, it is important to have basic standardized methods to culture stem cells. Such documented methods would allow for scientific reproducibility to enhance the efficient translation of stem-cell-based therapies to patients.

This inaugural volume on techniques to culture adult stem cells by River Publisher addresses this gap in the field of stem cells. Specifically, the volume is broadly divided first, with the basic methods to culture adult stem cells, followed by related methods to enhance the culture and characterization of stem cells. The key intent of this stem cell method book and future volumes is for laboratories across scientific entities – academic and commercial – to adapt the described methods when establishing cultures from various tissues. In light of the growing development of new technologies, these methods could serve as a starting platform that could be modified in different laboratories, depending on the scientific question. These changes would be considered when improving the current techniques in follow-up series.

Each chapter has a brief introduction of the specific stem cell, followed by a stepwise method with information on reagents. The chapter ends with notes to address potential pitfalls and other technical issues. Except for the chapter on iPSCs, the described techniques are specific to human stem cells. The techniques pertaining to human stem cell culture could be extrapolated when establishing similar techniques with animal tissues. As an example, the method to isolate hematopoietic stem cells, although similar, requires the use of different antibodies when the starting heterogeneous population of cells is from murine tissues such as femur.

The chapters were compiled by authors with expertise in the respective methods. The editor has decades of research with primary human adult stem cells, beginning with studies using hematopoietic stem cells. During her decades-long research career, the editor has developed methods to efficiently isolate various stem cells, which she has shared in this first volume.

The volume is intended for investigators at all levels – training and established scientists. The outlined methods provide detailed procedures to supplement the generally vague methods in peer-reviewed publications. The volume will benefit investigators who are planning to transition toward adult stem cell research and for those investigators who might not want stem cells to be the main emphasis within their group but rather to answer a scientific question that requires a few assays with adult stem cells.

The editor and River Publisher thank the authors and peer-reviewers who have spent time on this volume on stem cell techniques.

**Pranela Rameshwar**, Ph.D.
*Editor*

# List of Contributors

**Acevedo, Sebastian,** *Division of Hematology/Oncology, Department of Medicine, Rutgers New Jersey Medical School, USA*

**Ayer, Seda,** *Department of Medicine, Rutgers New Jersey Medical School, USA*

**Bujko, Kamila,** *Laboratory of Regenerative Medicine, Center for Preclinical Studies, Medical University of Warsaw, Poland*

**Chumak, Vira,** *Laboratory of Regenerative Medicine, Center for Preclinical Studies, Medical University of Warsaw, Poland*

**Einstein, Ella,** *Division of Hematology/Oncology, Department of Medicine, New Jersey Medical School, Rutgers Biomedical and Health Sciences, USA*

**Eljarrah, Adam,** *Department of Medicine, New Jersey Medical School, USA*

**Etchegaray, Jean-Pierre,** *Department of Biological Sciences, Rutgers University, USA*

**Fernandez-Moure, Joseph S.,** *Division of Trauma, Acute, and Critical Care Surgery, Department of Surgery, Duke University School of Medicine, USA*

**Ferrer, Alejandra I.,** *Department of Medicine, Rutgers New Jersey Medical School, USA; Rutgers School of Graduate Studies at New Jersey Medical School, USA*

**Gonzalez, Edward A.,** *Department of Biological Sciences, Rutgers University, USA*

**Greco, Steven S.,** *Department of Medicine, Rutgers New Jersey Medical School, USA*

**Guiro, Khadidiatou,** *Department of Medicine, Rutgers New Jersey Medical School, USA*

**Haddadin, Michael,** *Department of Medicine, Division of Hematology/Oncology, University of Massachusetts Chan Medical School, USA*

**Hudson, Taylor,** *Division of Trauma, Acute, and Critical Care Surgery, Department of Surgery, Duke University School of Medicine, USA*

**Kenfack, Yannick,** *Division of Hematology/Oncology, Department of Medicine, Rutgers New Jersey Medical School, USA; Rutgers School of Graduate Studies at New Jersey Medical School, USA*

**Kra, Joshua,** *Department of Medicine, Division of Hematology/Oncology, Rutgers New Jersey Medical School, Rutgers Cancer Institute of New Jersey at University Hospital, USA*

**Krishnamoorthy, K.,** *Department of Obstetrics, Gynecology and Women's Health, Rutgers New Jersey Medical School, Rutgers Biomedical and Health Sciences, USA*

**Kucia, Magdalena,** *Laboratory of Regenerative Medicine, Center for Preclinical Studies, Medical University of Warsaw, Poland*

**Lee, Edward S.,** *Department of Surgery, Rutgers New Jersey Medical School, USA*

**McMillian, Najerie,** *Division of Trauma, Acute, and Critical Care Surgery, Department of Surgery, Duke University School of Medicine, USA*

**Patel, Shyam A.,** *Department of Medicine, Division of Hematology/Oncology, University of Massachusetts Chan Medical School, USA*

**Perricho, Stephanie,** *Division of Hematology/Oncology, Department of Medicine, New Jersey Medical School, Rutgers Biomedical and Health Sciences, USA*

**Petryna, Andrew,** *Department of Medicine, Rutgers New Jersey Medical School, USA; Rutgers School of Graduate Studies at New Jersey Medical School, USA*

**Powell, K.,** *Department of Obstetrics, Gynecology and Women's Health, Rutgers New Jersey Medical School, Rutgers Biomedical and Health Sciences, USA*

**Rameshwar, Pranela,** *Division of Hematology/Oncology, Department of Medicine, Rutgers New Jersey Medical School, USA*

**Ratajczak, Mariusz Z.,** *Laboratory of Regenerative Medicine, Center for Preclinical Studies, Medical University of Warsaw, Poland; Stem Cell Program, University of Louisville, USA*

**Rojas, Darling,** *Rutgers School of Dental Medicine, USA*

**Romagano, Matthew P.,** *Department of Obstetrics, Gynecology and Women's Health, Rutgers New Jersey Medical School, Rutgers Biomedical and Health Sciences, USA*

**Rosemond, Harrison C.,** *Rutgers School of Graduate Studies at New Jersey Medical School, USA*

**Sandiford, Oleta A.,** *Department of Medicine, Rutgers New Jersey Medical School, USA*

**Sarkar, Anushka,** *Division of Hematology/Oncology, Department of Medicine, Rutgers New Jersey Medical School, USA; Rutgers School of Graduate Studies at New Jersey Medical School, USA*

**Savanur, Vibha Harindra,** *Department of Medicine, Rutgers New Jersey Medical School, USA; Rutgers School of Graduate Studies at New Jersey Medical School, USA*

**Shadpoor, Bobak,** *Division of Hematology/Oncology, Department of Medicine, Rutgers New Jersey Medical School, USA; Rutgers School of Graduate Studies at New Jersey Medical School, USA*

**Sherman, Lauren S.,** *Department of Medicine, Rutgers New Jersey Medical School, USA*

**Sinha, Garima,** *Department of Medicine, Rutgers New Jersey Medical School, USA; Rutgers School of Graduate Studies at New Jersey Medical School, USA*

**Souyah, Sami,** *Division of Hematology/Oncology, Department of Medicine, New Jersey Medical School, Rutgers Biomedical and Health Sciences, USA*

**Tsolaki, Ioanna,** *Department of Periodontics, Rutgers School of Dental Medicine, USA; Rutgers School of Graduate Studies at New Jersey Medical School, USA*

**Walker, Nykia D.,** *Department of Medicine, Rutgers New Jersey Medical School, USA; Rutgers School of Graduate Studies at New Jersey Medical School, USA*

**Wang, Dahui,** *Department of Biological Sciences, Rutgers University, USA*

**Williams, Shauna F.,** *Department of Obstetrics, Gynecology and Women's Health, Rutgers New Jersey Medical School, Rutgers Biomedical and Health Sciences, USA*

**Woloszyn, Derek J.,** *Department of Surgery, Rutgers New Jersey Medical School, USA*

# List of Figures

# List of Abbreviations

| | |
|---|---|
| **2i** | Two inhibitor |
| **5caC** | 5-carboxylcytosine |
| **5fC** | 5-formylcytosine |
| **5hmC** | 5-hydroxymethylcytosine |
| **5mC** | 5-methylcytosine |
| **ALP** | Alkaline phosphatase |
| **AM** | Amniotic membrane |
| **AML** | Acute myeloid leukemia |
| **AP** | Alkaline phosphatase |
| **APC** | Allophycocyanin |
| **APC7** | APC-Cy7 |
| **ASC** | Adipose-derived stem cells |
| **BC** | Breast cancer |
| **BCC** | Breast cancer cell |
| **bFGF** | Basic fibroblast growth factor |
| **BM** | Bone marrow |
| **BMSC** | Bone marrow stromal cell |
| **BSA** | Bovine serum albumin |
| **C/EBP$\alpha$** | CCAAT/enhancer binding protein-$\alpha$ |
| **CB** | Compact bone |
| **CFU** | Colony forming unit-fibroblast |
| **CM** | Chorionic membrane |
| **CPD** | Citric phosphate with dextrose buffer |
| **CSC** | Cancer stem cell |
| **CV** | Chorionic villi |
| **D** | Decidua |
| **DAPI** | 4′,6-diamidino-2-phenylindole |
| **DMEM** | Dulbecco's Modified Eagle Medium |
| **DMR** | Differentially methylated region |
| **DMSO** | Dimethyl sulfoxide |
| **Dox** | doxycycline |
| **DPBS** | Dulbeccos phosphate buffered saline |

| | |
|---|---|
| **DPSC** | Dental pulp stem cell |
| **EDTA** | Ethylenediaminetetraacetic acid |
| **ERK** | Extracellular signal-regulated kinase |
| **ESC** | Embryonic stem cell |
| **FACS** | Fluorescence-activated cell sorting |
| **FBS** | Fetal bovine serum |
| **FCS** | Fetal Calf Serum |
| **FGF4** | Fibroblast growth factor 4 |
| **FITC** | Fluorescein isothiocyanate |
| **GBM** | Glioblastoma multiforme |
| **G-CSF** | Granulocyte colony stimulating factor |
| **GFAP** | Glial fibrillary acidic protein |
| **GMP** | Good manufacturing procedure |
| **HBSS** | Hanks Balanced Salt Solution |
| **HS** | Horse sera |
| **HSA** | Human serum albumin |
| **HSC** | Hematopoietic stem cell |
| **HSPC** | Hematopoietic stem and progenitor cells |
| **hUCB** | Human umbilical cord blood |
| **HUVEC** | Human umbilical vascular endothelial cell |
| **IACUC** | Institutional Animal Care and Use Committee |
| **iPSC** | Induced pluripotent stem cell |
| **IRB** | Institutional Review Board |
| **ISCT** | International Society for Cellular Therapy |
| **JmjC** | Jumonji |
| **KG** | Ketoglutarate |
| **LAIP** | Leukemia-associated immunophenotype |
| **LIF** | Leukemia inhibitor factor |
| **LSC** | Leukemia stem cell |
| **mAbs** | Monoclonal antibodies |
| **MEF** | Mouse embryonic fibroblast |
| **MEF** | Murine embryonic fibroblast |
| **MEK** | Mitogen-activated protein kinase kinase |
| **MEM** | Minimal essential medium |
| **MFC** | Multiparameter flow cytometry |
| **miRNA** | microRNA |
| **MNC** | Mononuclear cell |
| **MPB** | Mobilized peripheral blood |
| **MPBSC** | Mobilized peripheral blood stem cell |
| **MRD** | Measurable residual disease |

| | |
|---|---|
| **MSC** | Mesenchymal stem cell |
| **NGS** | Next-generation sequencing |
| **NIM** | Neuronal induction medium |
| **NPC** | Neural progenitor cell |
| **NSC** | Neural stem cells |
| **PB** | Pacific blue |
| **PB** | Peripheral blood |
| **PBS** | Phosphate buffered saline |
| **PC7** | PE-Cy7 |
| **PCP5.5** | PerCP-Cy5.5 |
| **PE** | Phycoerythrin |
| **PE** | Preeclampsia |
| **P-MSC** | Placental stem cells |
| **P-S** | Penicillin–streptomycin |
| **PSC** | Pluripotent stem cells |
| **PVDF** | polyvinylidene difluoride |
| **RA** | Retinoic acid |
| **RBC** | Red blood cell |
| **RPMI** | Roswell Parker Memorial Institute |
| **RT** | Room temperature |
| **RT-PCR** | Reverse-transcriptase polymerase chain reaction |
| **SBA** | Soynean Agglutin |
| **SLAM** | CD150 |
| **SORE** | Sox2-Oct4a response element |
| **SSC** | Side scatter |
| **SVF** | Stromal vascular fraction |
| **Tet** | Ten–eleven translocation |
| **UC** | Umbilical cord |
| **UCB** | Umbilical cord blood |
| **VSEL** | Very Small Embryonic Like Cells |
| **WBC** | White blood cell |

# Introduction

## Core Techniques in Stem Cell Culture

**Pranela Rameshwar**

Department of Medicine, Rutgers New Jersey Medical School, USA
**Corresponding Author:** Pranela Rameshwar, Department of Medicine –
Division of Hematology/Oncology, Rutgers, New Jersey Medical School,
USA
Email: rameshwa@njms.rutgers.edu

## Abstract

The present collection of stem cell culture methods is the first volume for
a planned series on the details, limitations, and state-of-the-art solutions of
routine experimental methods to advance the area of stem cell research. Here
we attempted to standardize a given method(s) across laboratories while
gaining insights into the reason why we select particular stem cells in the
volume. The area of tissue regeneration and its emerging utility in treatments
will need to be specific to the source of stem cells and/or their specific secretome. The chapters provide culture method and associated information to
provide insights when developing hypotheses and a step-by-step guide on
testing the questions. Except for the method on induced pluripotent stem
cells (iPSC), the methods are focused on human tissue with the choice to
extrapolate the methodology to animal tissues. The editor hopes that both
biologists and non-biologists will be able to culture stem cells with ease.
This collection of methods will encourage large teams to reproduce other
results, work together, and to hasten clinical application. The primary focus
to culture stem cells is followed by other supporting methods required to
effectively culture the stem cells.

## Introduction

This first volume is for a planned series that pertains to different experimental
methods. The series begins with this volume of stem-cell-related methods.

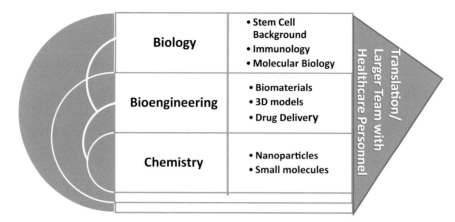

**Figure 1**   Progression to interdisciplinary stem cell research studies. Shown are examples of investigators of broad scientific disciplines (left boxes). The right boxes show how each discipline could contribute to broad areas of research to contribute to stem cell research. The connecting arrows provide insights into how the different disciplines could be linked for the translation of science.

The collection includes methods pertaining to most reported stem cells in the literature. A choice of selecting a given stem cell to include in the volume was also made based on the literature describing the translational potential. This book is intended to facilitate biologists and non-biologists with ease to incorporate stem cells in their research studies. The methods will encourage investigators with different scientific backgrounds to test hypotheses with stem cells in their model systems (Figure 1). The individual studies will provide information useful for allied health sciences' disciplines and other fields including material science and engineering. This would encourage teams of investigators to develop inter- and multi-disciplinary studies, which are needed to effectively translate stem cells in various areas of medicine. Together, these broad and novel approaches will encourage the establishment of larger teams with other expertise including clinician scientists. Consequently, the large teams will enhance the formulation of protocols to translate stem cells to patients (Figure 1). Indeed, stem cell research has learnt the need for inter-disciplinary teams [1]. We propose that a detailed documentation on stem cell culture and related methods would encourage investigators, who would otherwise find it difficult, and at times, intimidating when planning to venture into in-depth stem cell research. Consequently, these studies would enhance the translation of stem cells to patients to improve healthcare. The ongoing inclusion of engineers into the field of stem cells will also encourage parallel development of medical devices. The following sections provide an overview of the categories of methods while describing how the specific technique relates to the overall research on stem cell.

## Adult Stem Cell

This volume focuses on the isolation and expansion of selected adult human stem cells for the purpose of standardization across laboratories. The only exception is the method to isolate induced pluripotent stem cells (iPSCs), which describes a stepwise method of a known murine model. Investigators whose research involves animal models of stem cells could extrapolate the protocols using the literature to incorporate slight modifications. The majority of the stem cell culture techniques can be applied when culturing and isolating equivalent animal stem cells. The volume also includes key associated methods to provide ease when working with stem cells.

The first description includes what is considered as a key stem cell, namely, the isolation of hematopoietic stem cells (HSCs). These stem cells have been studied for more than a decade and formed the basis when studying other stem cells, in particular the role of cellular and other factors within the niche to regulate stem cells [2–4]. Thus, the isolation of HSCs could serve a frame of reference for concepts when studying other stem cells including those at low frequency [5].

The volume includes a chapter on very small embryonic-like cells (VSELs). These stem cells have been studied and have been shown to have translational potential [6]. VSELs were first described by Dr. Ratajczak, which was validated and reproduced by several independent laboratories [7]. VSELs can be expanded and have been shown to exhibit functional improvement of transplants [8]. Based on the promise of VESELs in regenerative medicine, the inclusion of this method could provide other scientists with information on these stem cells. This could begin by following the described method to culture VSELs.

Mesenchymal stem cells (MSCs), also referred to as mesenchymal stromal cells, are multipotent cells that are functionally plastic [9–11]. More importantly, MSCs show strong evidence that they can be used as off-the-shelf source in medicine. This special privilege of MSCs is mostly attributed to their ability to cross allogeneic barrier [12]. Indeed, the majority of clinical trials are conducted with MSCs and their secretome such as exosomes [13, 14] (ClinicalTrials.gov). MSCs can be expanded from different tissues such as adipose, placenta, and the Wharton Jelly of umbilical cord. Thus, the volume includes methods to culture MSCs from an adult human bone marrow, adipose, umbilical cord, placenta, bone chips, and dental pulp. An evolving information on MSCs indicates that a particular clinical application would require stem cells from a particular source. Therefore, the contributors addressed this by including a chapter on MSCs from the endosteal region, which has been shown to have a preference for osteogenic cells [15].

The large number of retracted papers on cardiac stem cells has been detrimental to the field [16]. Considering cardiac failure remains a clinical issues, the negative information on cardiac stem cells have slowed the progress in the field. Instead of halting the area of research, it is necessary to continue research studies. Although this volume did not include a method on cardiac stem cells, we envision that the outlined methods for other stem cells could serve as a basis to develop improved methods to determine the existence of cardiac stem cells. Perhaps such attempts might explain the reported increase in cardiomyocytc. Indeed, there is vast literature that indicates the potential of engineering techniques for cardiac repair, such as the incorporation of biomaterials and other methods [17].

The inclusion of iPSCs is outlined for a specific murine model. This chapter, similar to others in the volume, is prepared by an expert in the field. Despite the similarity with embryonic stem cells, iPSCs are considered adult stem cells since the source of IPSC would be selecting an adult cells to undergo dedifferentiation.

## Fetal Stem Cell

This section includes culture of two fetal stem cells – placenta and umbilical cord. Both tissue sources are discarded, which reduced the ethical concerns of using the derived cells. Both of these sources of stem cells have regenerative potential with a clear path in medicine [18, 19]. Recently, placenta stem cells have been associated with the pathology of preeclampsia [20]. Thus, in addition to its direct application in medicine, the dysfunction of placental stem cells in pregnancy could lead to the development of treatment.

## Cancer Stem Cell (CSC)

The volume includes the significance of CSCs in solid and one hematological cancer. The descriptions can be extrapolated for the isolation of any source of CSCs. The studies with solid tumors described the knowledge of investigators who isolated CSCs from breast cancer and glioblastoma as two examples of starting tumor tissues. These methods are important because the literature still lacks clear validated markers for CSCs. The method for solid tumors described a generalized protocol linked to stem cell regulation. These methods can be used, as described in the chapter. However, the methods on CSCs need to improve, perhaps applying the included chapter as a starting point.

## Other Supporting Methods

The isolation of HSCs and bone-marrow-derived MSCs (Methods 1 and 3) require material that needs to be procured through the donation of bone marrow aspiration. This will entail scientists to follow ethical regulation and interface with clinical personnel. This procedure is included at the beginning of the section. Studies of HSC function require long-term culture initiating cell culture. These cultures will require feeder stromal cells. The preparation of stromal culture needs human bone marrow aspirate and is therefore described in-depth. Culturing of stroma and HSCs with bone marrow aspirate will require the elimination of red blood cells. This method requires density gradient separation with Ficoll-Hypaque. This method is described in detail in the supporting method section.

Most, but not all of the methods will require characterization as part of standardization protocol. One such method is phenotyping by flow cytometry. Therefore, the supporting method section includes a generalized procedure that describes labeling and analyses of data from flow cytometry. Finally, since human tissues are limited, excess isolated stem cells should be cryopreserved. Thus, the final supporting method include details of saving stem cells for later use.

## References

[1] Stephens N, Khan I and Errington R. Analysing the role of virtualisation and visualisation on interdisciplinary knowledge exchange in stem cell research processes. Palgrave Commun. 2018; 4(1):78.

[2] Granot N and Storb R. History of hematopoietic cell transplantation: challenges and progress. Haematologica. 2020; 105(12):2716–2729.

[3] Nucifora G, Laricchia-Robbio L and Senyuk V. EVI1 and hematopoietic disorders: History and perspectives. Gene. 2006; 368:1–11.

[4] Qiu J, Papatsenko D, Niu X, Schaniel C and Moore K. Divisional History and Hematopoietic Stem Cell Function during Homeostasis. Stem Cell Reports. 2014; 2(4):473–490.

[5] Morrison SJ, Qian D, Jerabek L, Thiel BA, Park I-K, Ford PS, Kiel MJ, Schork NJ, Weissman IL and Clarke MF. A Genetic Determinant That Specifically Regulates the Frequency of Hematopoietic Stem Cells1. J Immunol. 2002; 168(2):635–642.

[6] Liu G, David BT, Trawczynski M and Fessler RG. Advances in Pluripotent Stem Cells: History, Mechanisms, Technologies, and Applications. Stem Cell Reviews and Reports. 2020; 16(1):3–32.

[7] Ratajczak MZ, Ratajczak J and Kucia M. Very small embryonic-like stem cells (VSELs) an update and future directions. Circulation Res. 2019; 124(2):208–210.

[8] Zuba-Surma EK, Guo Y, Taher H, Sanganalmath SK, Hunt G, Vincent RJ, Kucia M, Abdel-Latif A, Tang X-L, Ratajczak MZ, Dawn B and Bolli R. Transplantation of expanded bone marrow-derived very small embryonic-like stem cells (VSEL-SCs) improves left ventricular function and remodelling after myocardial infarction. J Cell Mol Med. 2011; 15(6):1319–1328.

[9] Sherman LS, Shaker M, Mariotti V and Rameshwar P. Mesenchymal stromal/stem cells in drug therapy: New perspective. Cytotherapy. 2017; 19(1):19–27.

[10] Greco SJ and Rameshwar P. Mesenchymal stem cells in drug/gene delivery: implications for cell therapy. Ther Deliv. 2012; 3(8):997–1004.

[11] Cho KJ, Trzaska KA, Greco SJ, McArdle J, Wang FS, Ye JH and Rameshwar P. Neurons Derived From Human Mesenchymal Stem Cells Show Synaptic Transmission and Can Be Induced to Produce the Neurotransmitter Substance P by Interleukin–1α. Stem Cells. 2005; 23(3):383–391.

[12] Potian JA, Aviv H, Ponzio NM, Harrison JS and Rameshwar P. Veto-Like Activity of Mesenchymal Stem Cells: Functional Discrimination Between Cellular Responses to Alloantigens and Recall Antigens1. J Immunol. 2003; 171(7):3426–3434.

[13] Squillaro T, Peluso G and Galderisi U. Clinical Trials with Mesenchymal Stem Cells: An Update. Cell Transplantation. 2016; 25(5):829–848.

[14] Lee B-C, Kang I and Yu K-R. (2021). Therapeutic Features and Updated Clinical Trials of Mesenchymal Stem Cell (MSC)-Derived Exosomes. J Clin Med. 2021, 10(4), 711.

[15] Fernandez-Moure J, Moore CA, Kim K, Karim A, Smith K, Barbosa Z, Van Eps J, Rameshwar P and Weiner B. Novel therapeutic strategies for degenerative disc disease: Review of cell biology and intervertebral disc cell therapy. SAGE Open Med. 2018; 6:1–11.

[16] Davis DR. Cardiac stem cells in the post-Anversa era. Eur Heart J. 2019; 40(13):1039–1041.

[17] Li Y, Wei L, Lan L, Gao Y, Zhang Q, Dawit H, Mao J, Guo L, Shen L and Wang L. Conductive biomaterials for cardiac repair: A review. Acta Biomaterialia. 2022; 139:157–178.

[18] McIntyre JA, Jones IA, Danilkovich A and Vangsness CT. The Placenta: Applications in Orthopaedic Sports Medicine. Am J Sports Med. 2017; 46(1):234–247.

[19] Fu-jiang CAO and Shi-qing F. Human umbilical cord mesenchymal stem cells and the treatment of spinal cord injury. Chinese Med J. 2009; 122(02):225–231.

[20] Romagano MP, Sherman LS, Shadpoor B, El-Far M, Souayah S, Pamarthi SH, Kra J, Hood-Nehra A, Etchegaray J-P, Williams SF and Rameshwar P. Aspirin-Mediated Reset of Preeclamptic Placental Stem Cell Transcriptome – Implication for Stabilized Placental Function. Stem Cell Rev Reports. 2022; 18(8):3066–3082.

# 1

## Isolation of Hematopoietic Stem Cells

**Bobak Shadpoor[1,2], Lauren S. Sherman[1,2], Sebastian Acevedo[1],**
**Yannick Kenfack[1,2], and Pranela Rameshwar[1]**

[1]Division of Hematology/Oncology, Department of Medicine, Rutgers New
Jersey Medical School, USA
[2]Rutgers School of Graduate Studies at New Jersey Medical School, USA
**Corresponding Author:** Pranela Rameshwar, Department of Medicine –
Division of Hematology/Oncology, Rutgers New Jersey Medical School,
USA
E-mail: rameshwa@njms.rutgers.edu
**Disclaimer:** The authors have nothing to declare.

## Abstract

Hematopoietic stem cells (HSCs) are the source of immune and blood cells.
In adults, the process occurs in the bone marrow, referred as hematopoietic
activity. The HSC role in hematopoietic activity has been studied for almost
a century; yet, to date, there is no documented method to expand these cells
for large numbers in experimental questions. The method to achieve a large
number of HSCs and their progenitors is fundamental to understand how
dysfunctions in aging can influence hematological diseases. This chapter
describes how to isolate human HSCs based on the current literature on the
phenotype of these cells. The described method is used for mobilized peripheral blood (MPB), BM aspirate and umbilical cord blood (UCB), and, if indicated, peripheral blood.

## 1.1 Introduction

Hematological malignancies, such as multiple myeloma, aplastic anemia, and
acute myeloid leukemia, are considered as aging disorders with disruptions
in the hematopoietic mechanisms [1]. Hematopoietic stem cells (HSCs) are

at the apex of the hematopoietic hierarchy, which gives rise to myeloid and lymphoid progenitors [2]. These processes are regulated by a complex hematopoietic microenvironment, consisting of cellular secretomes, to maintain the immune and blood systems throughout life [3–5]. Consequently, hematological disease typically manifests when there is a defect in the processes controlling the hematopoietic system.

Despite the hematopoietic system being studied for almost a century, the expansion of HSCs is still not a viable option for science. Thus, studies rely on the low frequency of HSCs, which is an even bigger issue for those studying human HSCs. Thus, it is critical to standardize and optimize HSC-isolation techniques to improve our understanding of the hematopoietic system and the possible sources of dysregulation. Such isolation can allow for improving on hematopoietic lineage development in the absence of a stress model with studies in radiated mice [6]. The isolation can also be used for more advanced technology like single-cell sequencing to determine hematopoietic cell heterogeneity and to determine if the available data regarding phenotype can predict the homogeneity of the isolated population.

## 1.2 Materials

*Introduction:* The described method allows investigators to use a method that is economically feasible – namely the use of SBA. The choice will depend on the scientific question. As an example, the experimental question might not require a highly purified CD34$^+$ population. In this case, one can use the SBA method.

1.   CD34 Microbead Human Kit (Miltenyi, Auburn, CA) or other commercial sources. Depending on the manufacturer, the instruction could vary.

2.   Mobilized peripheral blood (MPB), peripheral blood (PB), or umbilical cord blood (UCB) samples.

   a. Reference Note 1.

## 1.3 Method

1.   Obtain mobilized peripheral blood (MPB), umbilical cord blood (UCB), or bone marrow aspirate.

   a. Depending on the study, inclusion or exclusion criteria should be established when selecting the donor.

i. Regulatory approval: The study must be approved by your Institutional Review Board. This board will determine if there is a need for additional documentation, such as informed consent.

2. Separate the mononuclear cells (MNCs) by Ficoll Hypaque density gradient (Note 2).

3. SBA-depletion: Differentiated cells express carbohydrate N-acetyl glucosamine that binds to SBA. Deplete these cells SBA as follows [7]:

   a. Resuspend MNCs @ $2 \times 10^8$/ml in HBSS containing 2 µg/ml SBA (Note 3).

   b. Gently swirl the cell suspension until agglutination is visible.

   c. Layer suspension on 1 ml BSA solution. The agglutinated cells will settle by gravity after ~30 minutes.

   d. The suspension cells will include the CD34$^+$ cells and will be at the interface. Collect these cells and transfer in 1 vial containing 20 µl of 0.2 M galactose.

      i. Incubate on ice for ~15 minutes. This step will remove the SBA.

      ii. Wash 2× with HBSS.

4. Resuspend cells in a medium containing 5% FCS @ $5 \times 10^6$/ml. Add anti-human CD34 at final dilution = 1/50 (Note 4).

5. Incubate @ 4 °C for 4–6 hours, while rotating tubes with a rotating wheel – two-way mixer (Robbins Scientific).

6. Wash cells with a cold wash medium consisting of 2% FCS.

7. Resuspend cells @ $10^7$/ml and then proceed with the instruction from the isolation kit from any commercial source such as Miltenyi CD34 Microbead.

   a. Use the following rule of thumb to determine the quantity necessary: 200 µl volume of the beads/10 ml cell suspension.

8. Rotate the cells at @ 4 °C for 4–6 hours.

9. Select CD34$^+$ cells by magnetic separation (Notes 5 and 6).

## 1.4 Notes

Note 1: If necessary, researchers should reach out to their institution's Institutional Review Board (IRB) and gain permission to work with patient samples.

Note 2: Ficoll Hypaque density gradient protocol (see method in this book series).

    a. When isolating mononuclear cells from the buffy coat, attempt to minimize the amount of Ficoll you take since this will increase the density of the washing media and, at the same time, loss of cells in the media.

    b. Try to remove the buffy coat immediately after centrifugation because the prolonged presence of the cells in contact with Ficoll can cause cellular toxicity.

Note 3: In our lab, the amount of BM cells that we usually have will need 0.1–0.2 ml HBSS for this concentration of cell suspension.

    a. Use 1 µl of 100× SBA/0.1 ml cell suspension until agglutination is visible. This volume should be negligible and will not change the cell suspension or affect the concentration of balanced salt concentration.

Note 4: Researchers can also use anti-CD34 from alternative sources.

    a. Dynabeads CD34 Positive Isolation Kit.

    b. EasySep Human CD34 Positive Selection Kit II.

Note 5: When isolating CD34$^+$ HSPCs from blood samples, the yield will be low if the donor is healthy. Thus, in order to assess purity, employ immunochemical staining with an anti-CD34 rather than flow cytometry, which will require a large number of cells.

    a. You might be able to use flow cytometry if the isolation is from bone marrow aspirates. Again, this will depend on the volume of aspirate.

Note 6: One should be careful when labeling CD34$^+$ cells as HSCs because this marker is also expressed on hematopoietic progenitors and endothelial progenitors. The latter could be eliminated by labeling the isolated population for CD31.

b. Additionally, researchers could utilize other surface markers, such as SLAM protein surface markers CD150$^+$, CD48$^-$, and CD41$^-$ to identify HSCs [8].

## References

[1] Shlush LI. Age-related clonal hematopoiesis. Blood. 2018; 131(5):496–504.

[2] Grinenko T, Eugster A, Thielecke L, Ramasz B, Krüger A, Dietz S, Glauche I, Gerbaulet A, von Bonin M, Basak O, Clevers H, Chavakis T and Wielockx B. Hematopoietic stem cells can differentiate into restricted myeloid progenitors before cell division in mice. Nature Commun. 2018; 9(1):1898.

[3] Guidi N, Sacma M, Ständker L, Soller K, Marka G, Eiwen K, Weiss JM, Kirchhoff F, Weil T, Cancelas JA, Florian MC and Geiger H. Osteopontin attenuates aging-associated phenotypes of hematopoietic stem cells. EMBO J. 2017; 36(7):840–853.

[4] Kovtonyuk LV, Fritsch K, Feng X, Manz MG and Takizawa H. Inflamm-Aging of Hematopoiesis, Hematopoietic Stem Cells, and the Bone Marrow Microenvironment. Frontiers Immunol. 2016; 7.

[5] Pang WW, Schrier SL and Weissman IL. Age-associated changes in human hematopoietic stem cells. Semin Hematol. 2017; 54(1):39–42.

[6] Adigbli G, Hua P, Uchiyama M, Roberts I, Hester J, Watt SM and Issa F. Development of LT-HSC-Reconstituted Non-Irradiated NBSGW Mice for the Study of Human Hematopoiesis In Vivo. Front Immunol. 2021; 12:642198.

[7] Smith C, Gasparetto C, Collins N, Gillio A, Muench MO, O'Reilly RJ and Moore MA. Purification and partial characterization of a human hematopoietic precursor population. Blood. 1991; 77(10):2122–2128.

[8] Pinho S and Frenette PS. Haematopoietic stem cell activity and interactions with the niche. Nat Rev Mol Cell Biol. 2019; 20(5):303–320.

# 2

# Isolation Protocol for CD133⁺ and CD34⁺ Very Small Embryonic-like Cells (VSEL) from Human Umbilical Cord Blood

**Kamila Bujko¹, Vira Chumak¹, Mariusz Z. Ratajczak¹,², and Magdalena Kucia¹**

¹Laboratory of Regenerative Medicine, Center for Preclinical Studies, Medical University of Warsaw, Poland
²Stem Cell Program, University of Louisville, USA
**Corresponding Author:** Magdalena Kucia, Laboratory of Regenerative Medicine, Medical University of Warsaw
E-mail: magdalena.kucia@wum.edu.pl

## Abstract

The protocol presented here describes the procedures employed to identify and isolate very small embryonic-like stem cells (VSELs) from human umbilical cord blood (hUCB) derived samples using high-throughput fluorescence-activated cell sorting (FACS) machine. We thoroughly described the recommended steps to ensure their proper and efficient purification. VSELs are being investigated in over 20 laboratories worldwide; yet, this simple protocol was not always successfully transferred and reproduced. Therefore, here we focused on critical steps of great importance for the successful isolation to ensure the purity of the sorted population of VSELs.

## 2.1 Introduction

Very small embryonic-like stem cells (VSELs) are a rare population of epiblast-derived pluripotent stem cells. VSELs originate from cells related to the germline and are deposited in developing organs during embryogenesis, where they play a role as a backup population for monopotent tissue-committed stem cells [1, 2]. VSELs are quiescent but are activated during stressful situations and mobilized into the circulation. The number of these cells decreases with age. The existence of VSELs and their specification across germ layers differentiation was subsequently confirmed by at least 20 other independent groups [3–8].

VSELs are small cells, corresponding in size to the cells in the inner cell mass of the blastocyst, and they measure ~3–5 µm in mice and ~4–6 µm in humans. Human VSELs are lineage negative (Lin⁻), do not express lymphocyte common antigen CD45 antigen (CD45⁻), and display a primitive morphology (high nuclear/cytoplasm ratio and the presence of euchromatin in nuclei). Their existence has been confirmed and reported in umbilical cord blood (UCB), mobilized peripheral blood (mPB), and gonads [9]. Identification and purification of these cells by FACS sorting requires a unique gating strategy, mainly because of their small size. As they are slightly smaller than erythrocytes, the standard flow cytometry-based sorting protocols exclude those cells as they are typically considered to be cellular debris.

Properly isolated VSELs express markers of pluripotent stem cells (PSCs) identified in the embryonic stage, such as SSEA-1, Oct-4, Nanog, Rex-1, Dppa3, and Rif-1 in mice cells and SSEA-4, Oct-4, and Nanog in human cells. Murine BM VSELs erase the paternally methylated imprints at regulatory differentially methylated regions (DMRs) within the *Igf2-H19* and *Rasgrf1* loci; however, they also hypermethylate the maternally methylated imprints at DMRs for the *Igf2* receptor (*Igf2R*), *Kcnq1-p57$^{KIP2}$*, and *Peg1* loci [10]. Because paternally expressed imprinted genes (*Igf2* and *Rasgrf1*) enhance embryonic growth and maternally expressed genes (*H19*, *p57$^{KIP2}$*, and *Igf2R*) inhibit cell proliferation, the unique genomic imprinting pattern observed in VSELs demonstrates the growth-repressive influence of imprinted genes on these cells [11, 12].

Small cells comprising the population of human VSELs were also purified from neonatal umbilical cord blood (UCB) and adult patient-mobilized peripheral blood (mPB). Human UCB and mPB are enriched in VSELs within a population of CD133⁺ Lin− CD45⁻ cells that co-express CXCR4, and some of them are also CD34⁺. Like their murine counterparts, human UCB and mPB VSELs also express Oct4 and Nanog transcription factors in nuclei and express SSEA-1 on the surface [12, 13].

The bone marrow (BM) facilitates the maintenance of hematopoietic and non-hematopoietic stem cells in the unipotent or pluripotent state [13]. Despite this evidence, BM is not the best source for VSEL isolation since bone marrow biopsy is traumatic and painful for the patient. In turn, UCB represents an unlimited source of stem cells mainly used for hematopoietic stem cell transplantation with a large accessibility and tolerance to allogenic graft [15–17]. The cells can be isolated after birth without any risk for both mother and the newborn [18]. Purified human UCB-VSELs are enriched for cells that express some markers characteristic of pluripotent stem cells (e.g., Oct-4, Nanog, and SSEA-4) and some epiblast and primordial germ cell markers.

## 2.2 Materials

### 2.2.1 Human umbilical cord blood (hUCB)

1. Prior to starting your assay, ensure that you have the Institutional Review Board (IRB) approval. The hUCB can be obtained through an agreement with a medical center as, in most cases, the hUCB is treated as a medical waste and recycled together with the placenta and the cord. Otherwise, you can purchase whole hUCB, hUCB-derived mononuclear cells (MNCs), or even enriched in $CD34^+$ cells. Regardless of the source, the hUCB should be collected during the labor of healthy full-term newborns with written consent from the donor.

2. The hUCB should be collected into a dedicated blood bag containing anticoagulant. Recommended anticoagulants are citric phosphate with dextrose buffer (CPD) or ethylenediaminetetraacetic acid (EDTA).

3. To increase the yield of isolated cells, hUCB should be processed for no longer than 24 hours after collection. Transportation of the hUCB to the laboratory at 4 °C is advised.

### 2.2.2 Reagents for hUCB CD34+ and CD133+ VSEL isolation

1. Ficoll-Paque PLUS (density 1.077 g/ml) (GE Healthcare, Cat. No. 17-1440-03 or equivalent).

2. Phosphate buffered saline pH 7.2 (PBS).

3. RPMI-1640 medium (supplemented with L-glutamine and glucose) with 2% of fetal bovine serum (FBS).

### 2.2.3 Antibodies

- Lineage (Lin) cocktail of antibodies (Abs) (all FITC-conjugated mouse Abs): anti-CD235a (clone GA-R2 [HIR2], Cat. No. 559943), anti-CD2 (clone RPA-2.10, Cat. No. 555326), anti-CD3 (clone UCHT1, Cat. No. 555332), anti-CD14 (clone M5E2, Cat. No. 555397), anti-CD16 (clone 3G8, Cat. No. 555406), anti-CD19 (clone HIB19, Cat. No. 555412), anti-CD24 (clone ML5, Cat. No. 555427), anti-CD56 (clone NCAM16.2, Cat. No. 345811), and anti-CD66b (clone G10F5, Cat. No. 555724) (all BD Biosciences).

- CD45: PE-conjugated, mouse anti-CD45 (clone HI-30, Cat. No. 555483) (BD Biosciences).

- CD133: APC conjugated, mouse anti-CD133 (clone AC133, Cat. No. 130-113-668) (Miltenyi Biotec).

- CD34: APC conjugated, mouse anti-CD34 (clone 581, Cat. No. 555824) (BD Biosciences).

- Flow Cytometry Size Calibration Kit compatible with the machine (Flow Cytometry Size Calibration Kit, Invitrogen, Cat. No. F13838).

### 2.2.4 Plastics and equipment

- Falcon 50 ml tubes, 40 μm mash cell filters, 5 ml cytometer tubes and collection tubes, serological 10 ml pipettes, and tips 1-1000 μl with appropriate pipets.

- Centrifuge with cooling and slow acceleration/deceleration mode.

- Cell sorter recommendation: use a high-speed flow cytometry sorter with a 70 μm nozzle tip.

## 2.3 Methods

### 2.3.1 Isolation of MNCs

1.  Dilute hUCB samples with PBS (0.5–1× – the volume of the anticoagulant). The proper dilution is vital as the aggregates of red blood cells might trap MNCs in their clumps.

2.  Carefully layer 35 ml of the diluted hUCB sample over 15 ml of Ficoll-Hypaque PLUS (density 1.077 g/ml, room temperature) in a 50 ml Falcon tube.

3.  Centrifuge at 4 °C for 30 minutes, 400 × g on the lowest acceleration/deceleration mode.

4.  Collect the upper layer containing plasma and platelets using a serological pipet and discard. Leave about 1 cm of upper plasma/platelet layer to ensure the intact of MNCs.

5. Collect the MNCs from the interface using a 1000 μl pipette and transfer them to a new 50 ml Falcon tube. The MNC layer should appear as a fluffy white coat.

6. Fill up the tube with 1640-RPMI with 2% FBS (collected MNCs should be diluted at least 3× to ensure proper washing).

7. Centrifuge at 4 °C for 15 minutes, 600× *g*. The break can be turned on.

8. Discard the supernatant and resuspend cells in 1640-RPMI with 2% FBS for counting.

9. Count the cells to estimate the final resuspension for staining.

10. Wash the cells in 20–25 ml of 1640-RPMI with 2% FBS and centrifuge at 4 °C for 15 minutes, 600× *g*.

11. Resuspend the cells in 1640-RPMI with 2% FBS. Volume should be recalculated based on cell number and concentrations of antibodies used in further steps.

## 2.3.2 Staining for CD133⁺ or CD34⁺ VSELs (see Notes 1 and 2)

1. Stain single-cell suspension (use 40 μm mash cell filters and mix thoroughly) with the manufacturer's recommended concentration of antibodies listed in the Section 2.2. Ensure you prepare appropriate unstained and single fluorochrome controls. Incubate for 30 on ice, in the dark. Mix occasionally.

2. Wash cell suspension in RPMI-1640 medium with 2% FBS and centrifuge at 4 °C for 15 minutes using 600× *g*. Recommended washing in 50 ml Falcon tube.

2.3.3. Resuspend cells in RPMI-1640 medium with 2% FBS and filter through 40 μm mash before sorting. Recommended suspension is 2–3 ml for $1 \times 10^6$ cells (do not exceed $5 \times 10^6$ cells per ml).

## 2.3.3 Sorting protocol

1. Set up the machine accordingly (see Note 3).

2. Prepare the protocol. The dot plots are a recommended data display (see Note 4).

3. Run your "unstained" control tube to set the FSC/SSC gate. Run the mixture of predefined-sized microspheres to set a proper gate. The

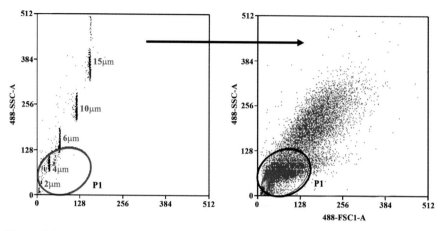

**Figure 2.1**  Gating strategy. Use of the predefined-sized microspheres for properly setting FSC/SSC gate to include events ranging 2–10 μm in size.

recommended size of the beads should range from 1 to 15 μm (see Note 5). Set up the gate to include only small events, in size of 2–10 μm, as shown in Figure 2.1. Further, analyze events from the P1 gate for the expression of the Lin marker.

4.  Next, set the gate to analyze the Lin marker expression. Ensure that the proper voltage is applied for the canal by running an "unstained" tube and, subsequently, run the sample stained with Lin antibodies only (FITC) to distinguish negative and positive events. Set gate P2 to include only Lin negative events; see Figure 2.2(A).

5.  Analyze Lin negative events from the P2 gate for the expression of CD45 and CD133 or CD34. Analyze the "unstained sample" to ensure that the proper voltage is applied and to differentiate negative events. Run CD45 (PE) only stained tube and subsequently analyze a CD34 (APC) stained tube to set gates. Set a gate P3 for the control population of HSC as Lin⁻CD45⁺CD34⁺ events and P4 for VSELs as Lin⁻ CD45⁻CD34⁺ events (Figure 2.2(B)); see Note 6.

6.  Run sample stained with a mix of antibodies. Ensure the proper flow rate recommended by the sorter machine. To avoid clogging, run a small sample volume at once and do not exceed recommended cell concentration. Sort the population of Lin⁻CD45⁻CD34⁺ VSELs and Lin⁻ CD45⁺CD34⁺ HSCs as shown in Figure 2.3.

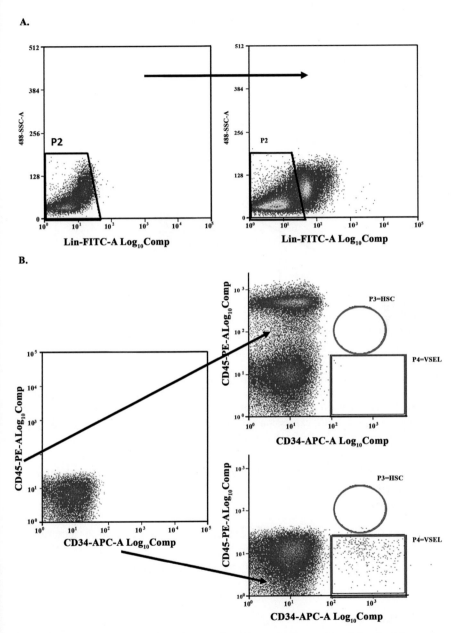

**Figure 2.2** Gating strategy. Cells from region P1 are further analyzed for the Lin marker expression (Lin) (Panel A) and CD45 (PE) and CD34 (APC) to set a gate for HSCs (Lin⁻CD45⁺CD34⁺) (P3) and VSELs (Lin⁻CD45⁻CD34⁺) (P4).

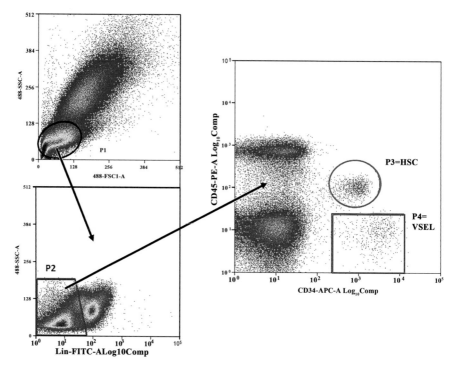

**Figure 2.3**    Gating strategy for sorting hUCB-derived VSELs and HSCs. hUCB-derived stained MNCs were visualized by dot plot presenting FSC/SSC signal. Small events, 2–10 μm in size (P1), were analyzed for the expression of Lin markers. Negative events from gate P2 were further analyzed for the expression of CD45 and CD34 antigens and sorted as the population of HSCs (Lin⁻CD45⁺CD34⁺) (P3) and VSELs (Lin⁻CD45⁻CD34⁺) (P4).

7.    Analyze the events from HSC and VSEL gates (P3 and P4, respectively) to verify the size of the events, as VSELs should be 2–6 μm in size; see Figure 2.4.

8.    The population of Lin⁻CD45⁻CD133⁺ VSELs can be sorted accordingly. Steps 3.1–3.5 are identical for both protocols, whereas Step 3.6 is different by using CD133 (APC) antibody to set the exact same gates (see Note 1).

## 2.4 Notes

Note 1: The described protocol utilizes the same fluorochrome Ab for CD34 and CD133 markers.

Therefore, they cannot be analyzed and sorted simultaneously. Should you consider sorting both populations, we recommend using the following mix of antibodies: Lin: unchanged from the above protocol; CD45: PE-Cy7

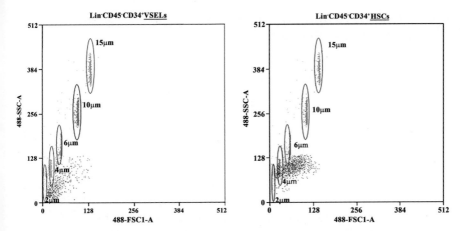

**Figure 2.4** Verification of the size of sorted cells. The sorted cells were down tracked to the P1 gate and aligned with the image of predefined-sized microspheres to verify the size of sorted VSELs and HSCs.

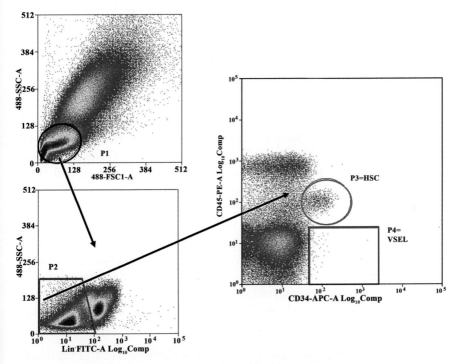

**Figure 2.5** Alternative sorting strategy for Lin⁻CD45⁻CD133⁺ VSELs. hUCB-derived stained MNCs visualized by dot plot presenting FSC/SSC signal. Small events, 2–10 μm in size (P1), were analyzed for the expression of Lin markers. Negative events from gate P2 were further analyzed for the expression of CD45 and CD133 antigens and sorted as the population of HSCs (Lin⁻CD45⁺CD133⁺) (P3) and VSELs (Lin⁻CD45⁻CD133⁺) (P4).

mouse anti-CD45 (BioLegend, Cat. No. 304016); CD133 unchanged; CD34: PE mouse anti-CD34 (BioLegend, Cat. No. 343506).

Note 2: Additional steps of magnetic pre-sorting could be applied to this method. Sorting from the whole hUCB after Ficoll density gradient centrifugation is time-consuming (see also Note 3); therefore, if isolating cells from more than 50 ml of entire fresh hUCB, the sample size could be reduced by employing magnetic pre-sorting. autoMacs Pro Separator or Macs columns for manual separation (Miltenyi Biotec) can be used after labeling with specific micro beads. There are few options to enrich the sample in the desired populations of cells. If both subpopulations of VSELs are being sorted, the Lineage Depletion Kit (Miltenyi Biotec, Cat. No. 130-092-211) could be used to deplete all mature hematopoietic cells and their committed precursors. If sorting the population of CD34⁺ VSELs, the CD34 MicroBead Kit (Miltenyi Biotec, Cat. No. 130-046-702) is recommended as it allows for the positive selection of CD34⁺ hematopoietic stem and progenitor cells. The CD133 MicroBead Kit – Hematopoietic Tissue can be used to enrich the sample if sorting CD133⁺ subpopulations of cells. If any of the above pre-sorting methods were used, the following steps of Ab staining and sorting would remind the same.

Note 3: The following protocol was established on MoFlo Astrios EQ, a high-speed flow cytometry sorter (Beckman Coulter). The machine enables >100,000 events/second acquisition rate with 70,000 validated sort decisions/sec rate, although we do not recommend exceeding 25,000–30,000 events/second maximum speed with the 70 μm nozzle tip. Note that the sorting machine can be equipped with differently sized nozzles, which will affect the sorting speed. Sorting VSELs and HSCs from the whole hUCB can be time-consuming as even small volumes of blood contain a high number of cells; therefore, the smallest possible (based on the size of sorted cells) nozzle tip should be used. Another thing that should be taken into consideration is the sort decision mode. The MoFlo Astrios EQ allows to choose from three different modes depending on the desired output. Even though the number of VSELs is not high, we recommend selecting a mode focused on purity rather than recovery. The proposed protocol requires two lasers: "blue" (488 nm) and "red" (340 nm).

Note 4: The VSELs population, even if sorted from enriched in stem cells source as an hUCB, is scarce; therefore, the flow cytometry analysis should be carried out using a dot plot to identify these cells correctly.

Note 5: Setting the proper FSC/SSC gate is the most critical step in sorting VSELs; therefore, the size beads approach is obligatory. The gate should be set to exclude contamination by cell debris, and small erythroblasts, as sorting undesired objects could have a significant impact on further functional experiments or analysis. Make sure the instrument threshold does not exclude small events. Of note, even though the size of the VSELs was confirmed by microscopic measurements and analysis, the resolution power and accuracy of the cell sorter machine should be considered as it might affect how the size beads are visualized. The signal generated by artificially made beads might also be slightly different than the one generated by same-sized living cells. Therefore, 2 μm buffer is recommended. For more advice on the matter of setting the FSC/SSC gate, see Note 6.

Note 6: This protocol was designed to sort VSELs efficiently. Although in our analysis, we also include HSC as the applied antibodies set enables us to detect the population of HSC as well. Simultaneously analyzing both populations allows us to verify the sample quality and efficiency of sorting. hUCB is enriched in the HSC population; therefore, we are able to detect a high number of these cells in our sample. Of note, both populations, VSELs and HSCs, are detected in a 2–10 μm gate, where those two populations slightly overlap in size as the "biggest" VSELs are of the size of the "smallest" HSC.

## References

[1] Leppik L, Sielatycka K, Henrich D, Han Z, Wang H, Eischen-Loges MJ, Oliveira KMC, Bhavsar MB, Ratajczak MZ, Barker JH. Role of Adult Tissue-Derived Pluripotent Stem Cells in Bone *Regeneration. Stem Cell Rev Rep*. 2020; 16(1):198–211.

[2] Sovalat H, Scrofani M, Eidenschenk A, Pasquet S, Rimelen V, Hénon P. Identification and isolation from either adult human bone marrow or G-CSF-mobilized peripheral blood of CD34(+)/CD133(+)/CXCR4(+)/ Lin(-)CD45(-) cells, featuring morphological, molecular, and phenotypic characteristics of very small embryonic-like (VSEL) stem cells. *Exp Hematol*. 2011; 39(4):495–505.

[3] Guerin CL, Loyer X, Vilar J, Cras A, Mirault T, Gaussem P, Silvestre JS, Smadja DM. Bone-marrow-derived very small embryonic-like stem cells in patients with critical leg ischaemia: evidence of vasculogenic potential. *Thromb Haemost*. 2015; 113(5):1084–94.

[4] Zhang Q, Yang YJ, Qian HY, Wang H, Xu H. Very small embryonic-like stem cells (VSELs)-a new promising candidate for use in cardiac regeneration. *Ageing Res Rev*. 2011;10(1):173–7.

[5] Kassmer SH, Jin H, Zhang PX, Bruscia EM, Heydari K, Lee JH, Kim CF, Kassmer SH, Krause DS. Very small embryonic-like stem cells from the

murine bone marrow differentiate into epithelial cells of the lung. *Stem Cells.* 2013; 31(12):2759–66.

[6] Wojakowski W, Ratajczak MZ, Tendera M. Mobilization of very small embryonic-like stem cells in acute coronary syndromes and stroke. *Herz.* 2010; 35(7):467–72.

[7] Bhartiya D, Kasiviswananthan S, Shaikh A. Cellular origin of testis-derived pluripotent stem cells: a case for very small embryonic-like stem cells. *Stem Cells Dev.* 2012; 21(5):670–4.

[8] Gharib SA, Dayyat EA, Khalyfa A, Kim J, Clair HB, Kucia M, Gozal D. Intermittent hypoxia mobilizes bone marrow-derived very small embryonic-like stem cells and activates developmental transcriptional programs in mice. *Sleep.* 2010 Nov;33(11):1439–46.

[9] Suszynska M, Zuba-Surma EK, Maj M, Mierzejewska K, Ratajczak J, Kucia M, Ratajczak MZ. The proper criteria for identification and sorting of very small embryonic-like stem cells, and some nomenclature issues. *Stem Cells Dev.* 2014; 23(7):702–13.

[10] Shin DM, Zuba-Surma EK, Wu W, Ratajczak J, Wysoczynski M, Ratajczak MZ, Kucia M. Novel epigenetic mechanisms that control pluripotency and quiescence of adult bone marrow-derived Oct4(+) very small embryonic-like stem cells. *Leukemia.* 2009; 23(11):2042–51.

[11] Kucia M, Masternak M, Liu R, Shin DM, Ratajczak J, Mierzejewska K, Spong A, Kopchick JJ, Bartke A, Ratajczak MZ. The negative effect of prolonged somatotrophic/insulin signaling on an adult bone marrow-residing population of pluripotent very small embryonic-like stem cells (VSELs). *Age (Dordr).* 2013; 35(2):315–30.

[12] Ratajczak MZ, Shin DM, Liu R, Mierzejewska K, Ratajczak J, Kucia M, Zuba-Surma EK. Very small embryonic/epiblast-like stem cells (VSELs) and their potential role in aging and organ rejuvenation--an update and comparison to other primitive small stem cells isolated from adult tissues. *Aging (Albany NY).* 2012; 4(4):235–46.

[13] Ratajczak MZ, Mierzejewska K, Ratajczak J, Kucia M. CD133 Expression Strongly Correlates with the Phenotype of Very Small Embryonic-/Epiblast-Like Stem Cells. *Adv Exp Med Biol.* 2013; 777:125–41.

[14] Shiozawa Y, Havens AM, Pienta KJ, Taichman RS. The bone marrow niche: habitat to hematopoietic and mesenchymal stem cells, and unwitting host to molecular parasites. *Leukemia.* 2008; 22(5):941–50.

[15] Gluckman E, E Luckma, HA Broxmeyer, AD Auerbach, HS Friedman, GW Douglas, A Devergie, H Esperou, D Thierry, G Socie, P Lehn, S Cooper, D English, J Kurtzberg, J Bard and EA Boyse. Hematopoietic

reconstitution in a patient with Fanconi's anemia by means of umbilical-cord blood from an HLA- identical sibling. *N Engl J Med.* (1989); 321(17):1174–8.

[16] Halasa M, Baskiewicz-Masiuk M, Dabkowska E, Machalinski B. An efficient two-step method to purify very small embryonic-like (VSEL) stem cells from umbilical cord blood (UCB). *Folia Histochem Cytobiol.* 2008; 46(2):239–43.

[17] Lahlil R, Scrofani M, Barbet R, Tancredi C, Aries A, Hénon P. VSELs Maintain their Pluripotency and Competence to Differentiate after Enhanced Ex Vivo Expansion. *Stem Cell Rev Rep.* 2018; 14(4):510–524.

[18] Monti M, Imberti B, Bianchi N, Pezzotta A, Morigi M, Del Fante C, Redi CA, Perotti C. A Novel Method for Isolation of Pluripotent Stem Cells from Human Umbilical Cord Blood. *Stem Cells Dev.* 2017;26(17):1258–1269.

# 3

# Isolation and Expansion of Mesenchymal Stem Cells from the Adult Bone Marrow

**Lauren S. Sherman[1,2], Harrison C. Rosemond[2], and Pranela Rameshwar[1]**

[1]Department of Medicine, Rutgers New Jersey Medical School, USA
[2]Rutgers School of Graduate Studies at New Jersey Medical School, USA
**Corresponding Author:** Pranela Rameshwar, Rutgers New Jersey Medical School, USA
E-mail: rameshwa@njms.rutgers.edu
**Disclaimer:** The authors have nothing to declare.

## Abstract

Mesenchymal stem cells (MSCs) can be derived from various tissues. There is compelling evidence that MSCs can be safely translated to patients without the hindrance of allogeneic match, termed off-the-shelf MSCs. Bone marrow (BM) aspirate was initially studied for MSCs and its utility to be licensed as immune suppressor cells. This was followed by exhaustive studies on other MSC sources. Thus, BM-derived MSCs could serve as a basic method when culturing MSCs from other tissues. This chapter outlines the details needed to isolate and expand human MSCs from bone marrow aspirates.

## 3.1 Introduction

Stem cells hold great promise for treating a variety of pathological conditions, ranging from drug delivery to acting as anti-inflammatory mediators, to regenerative capacity. Mesenchymal stem cells (MSCs) are a class of adult stem cells that can be isolated from numerous adult tissues including adipose, bone, bone marrow, dental pulp, and placenta, among others. Although derived from different tissues, the MSCs all share key characteristics including adherence to plastic; self-renewal, multilineage differentiation

21

capacity (e.g., adipocytes, osteocytes, and chondrocytes); ability to home in on areas of inflammation – including crossing the blood–brain barrier, capacity to be licensed or educated by an inflammatory microenvironment into immune-suppressive cells, and cell surface expression including CD34⁻, CD45⁻, HLA-DR⁻, CD44⁺, CD73⁺, CD90⁺, CD105⁺, and STRO-1⁺[1, 2]. Minor differences in phenotype and function have been attributed to MSCs derived from different tissue sources, but the key functions of MSCs as listed above are shared by MSCs derived from each source [3, 4].

Bone marrow (BM) derived MSCs are often preferable for research due to their ease of collection. BM aspirates can be collected from healthy subjects as an outpatient procedure. Additionally, MSCs can be obtained as excess discard tissue with bone marrow biopsy. However, if you opt to use this latter source of tissue, you must be cautious that these derived MSCs might be specialized for generating bone-forming cells [5]. Once the MSCs have been isolated from the broader BM milieu, the MSCs can be passaged and expanded up to tens of passages in culture. However, this does not mean you cannot continue passaging if the technique in a particular laboratory is careful.

## 3.2 Materials

*All materials should be sterile and used under aseptic conditions unless stated otherwise.*

1. Approval from your Institutional Review Board (IRB) is required. If you are getting BM aspirate from a commercial source, check with your IRB. You should also inform the institutional laboratory safety board that you plan to use human cells.

2. Dulbecco's Modified Eagle Medium (DMEM), with high glucose and sodium bicarbonate and without L-glutamine and sodium pyruvate (DMEM; Sigma Aldrich D5671 or other commercial source).

3. L-glutamine solution (200 mM).

4. Penicillin–streptomycin antibiotic solution (10,000 IU penicillin and 10,000 µg/ml streptomycin).

5. Fetal bovine serum, heat inactivated (FBS, Note 1).

6. 10 cc syringes (without needles).

7. Heparin solution (Note 2).

8. Bone marrow aspiration needle and accessories (see Method 13 for details).

9. 100 mm round, plasma-gas-treated tissue culture dishes (Falcon 353003 or other tissue culture dishes with plasma treated surfaces, Note 3).

10. Phosphate buffered saline (PBS).

11. 0.05% Trypsin-EDTA solution (trypsin)

12. 15 and/or 50 ml centrifuge tubes.

## 3.3 Methods

### 3.3.1 Bone marrow collection (also Method 13)

1. Preparation before obtaining the BM specimen:

    i. Prepare room temperature MSC media prior to obtaining the bone marrow aspirate. MSC media is composed of DMEM supplemented with 10% fetal bovine serum, 1% penicillin–streptomycin solution, and 1% L-glutamine.

    ii. Prepare 2–3 10 cc syringes with 100 units of heparin per syringe under aseptic conditions (Notes 2, 4, and 5).

    iii. Obtain informed consent from the bone marrow donor.

2. Bone marrow aspiration:

    i. A trained clinician will conduct the bone marrow aspirate, aspirating up to 10 ml of bone marrow per syringe (Note 6).

    ii. As collection of the syringe is complete or as soon as possible, invert the syringe several times to ensure the heparin mixes with the sample to prevent clotting. The mixed syringes should be returned to a "clean" transport bag (Note 5).

### 3.3.2 Plating bone marrow aspirates

1. Calculate the number of 100 mm tissue culture dishes needed: each dish will receive approximately 2 ml of aspirate.

2. Add 8 ml of MSC media to each tissue culture dish and label the lid.

3. Gently invert the syringes of BM to mix the cells and then gently extrude approximately 2 ml of aspirate, dropwise, into each plate.

4.   Gently swirl the plate to ensure the BM aspirate is equally spread across the plate's surface.

5.   Transfer the plated bone marrow aspirates to a 37 °C, 5% $CO_2$ incubator (Note 7).

### 3.3.3  Red blood cell removal

1.   After three days, the non-adherent fractions are collected from each dish and subjected to density gradient centrifugation to remove the red blood cells (Note 8; Chapter 15).
2.   Gently swirl the dishes to remove the maximum number of red blood cells within the non-adherent cells. If you opt to pipette up and down to remove the non-adherent cells that have weak adhesion onto the dish, be gentle.

3.   After removing the non-adherent cells, add 2–5 ml MSC media to each plate. This will avoid cell death due to drying of the surface.

4.   Return the recovered mononuclear cells to the culture dishes and then increase the total volume to 10 ml per dish with MSC media (Note 9).

### 3.3.4  Culturing of MSCs

1.   Replace 50% of the MSC media with fresh MSC media every 3–4 days until the cells reach 70%–80% confluence (Notes 10 and 11).

2.   At 70%–80% confluence, remove the media but collect in a sterile tube or bottle. If you plan to store the media as supplement for other MSC cultures, ensure that the stored media are cell-free. To do this, centrifuge at 500 $g$ and then store the supernatant (Note 15).

3.   Gently wash the dish with tissue-culture grade PBS. Gently swirl the plates or rock the dishes back and forth. Be careful that you do not spill the media or this could lead to contamination. Aspirate the PBS. This step is needed to remove residual MSC media that contain 10 FBS. Serum will prevent trypsin acting on the cells to de-adhere from the plastic.

4.   Disassociate MSCs by adding 1–2 ml trypsin in each 100 mm dish and then place the dishes in a 37 °C incubator for 5–10 min (Note 12).

5.   Examine the plates under an inverted microscope to determine if the MSCs have de-adhered. If not, replace the plates in the incubator. If the MSCs are disassociated, collect the trysin containing the MSCs and

transfer to an appropriate sized centrifuge tube containing media with 5% FBS. You may use any media since this will not be the incubation media. The FBS will dilute the trypsin to prevent further degradation of the cells (Note 13).

6.  Centrifuge the cells at 300 *g* at room temperature for 5–10 minutes.

7.  Aspirate the supernatant. Be careful as you approach the cell pellet to avoid aspirating the cells. Immediately break the pellet by tapping the tubes with your fingers. If you leave the cells in a pellet for a prolonged period, this would diminish the cell viability. Add MSC media to the cells. The total volume should allow you to accurately count the total number of cells in suspension. This could be gauged from the pellet size. If you are inexperienced with cell culture, it is better to add less media. At cell count, if there is too many cells in the hemacytometer (used for manual cell count – Method 18), add more media. If you have few cells in the hemacytometer, centrifuge the cells and remove excess media. After this, re-suspend the cell pellet.

8.  Count the cells manually (Method 18) or by an automated method in your laboratory. As a rule of thumb, seed 125,000–150,000 MSCs per 100 mm dish. Upon seeding the MSCs, increase the passage number *n* + 1 (Note 14). If you are an experienced laboratory, you may skip the counting procedure – transfer the total number of retrieved cells into three times the total number of plates. This will be reusing the original plate and then added two other plates.

9.  You may add MSC media to the plate before adding the cells or you may dilute the cells in the total amount of plates. The latter might not be practical because this will require large amount of media. The former method will require adding MSC media to the dishes and then adding the MSCs. It is better to dropwise add the MSCs for equal distribution across the plate. The media used should be a combination of previously used MSC media and fresh MSC media for a total volume of approximately 10 ml (Note 15).

10. Distribute the media on the dish by gently swirling. Place the dishes in the incubator.

11. Continue passaging the cells as described at 70%–80% confluence (approximately every 3–4 days).

12. By passage 3, the MSCs should be devoid of hematopoietic cells and fibroblasts. Begin to characterize the MSCs by phenotyping using flow

cytometry and multilineage differentiation (see Chapter 16 for MSC characterization).

13. Any passage you may cryopreserve the MSCs. It is particularly advisable to cryopreserve the early passaged cells at $5 \times 10^5$ to $1 \times 10^6$ cells/ 1 ml vial. Refer to Chapter 17 for the procedure.

## 3.4 Notes

Note 1: FBS must be batch-tested to ensure that it can support the growth of MSCs. The FBS should allow the cells to expand and to maintain multipotency. It is also good to compare with the current batch of FBS. If you are a new laboratory working with MSCs, use the standardized criteria to select the appropriate media.

Note 2: A stock heparin solution (e.g., 50–100 units/ml) can be made in DMEM and stored at 4 °C for several months.

Note 3: In our hands, the MSCs grow with better efficiency and longevity in tissue culture dishes as compared to tissue culture flasks.

Note 4: Collecting >25 ml of BM is discouraged since you are likely to have more blood than aspirate. If the healthcare provider prefers to collect the aspirate in syringes of a different volume, the heparin units can be scaled accordingly, e.g., 50 units of heparin in a 5 cc syringe.

Note 5: The syringes can be transported to the clinic in a "clean" rather than aseptic manner. We recommend returning the syringe to its sterile packaging and placing the syringes into a biohazard or specimen bag.

Note 6: This procedure of obtaining BM aspiration should be performed by a trained clinician. See Method 13 for a detailed procedure. It is preferable for the procedure to be conducted by a clinical hematologist who regularly preforms BM aspiration.

Note 7: Do not stack >4 dishes in the incubator since this could affect adequate airflow in the lower dishes.

Note 8: The red blood cells should be removed at as close to three days. If the red blood cells begin to lyse after day 3, the lysates will be toxic to the adherent cells.

Note 9: 100 ml dishes should have no less than 5 ml of media to prevent the media to dry during incubation. No more than 12–13 ml should be used to ensure adequate gas exchange at the dish's base.

Note 10: Do not permit the MSCs to reach greater than 70%–80% confluency. Upon contact, the MSCs will begin to differentiate. Avoid passaging at <70% confluence. In our hands, this diminishes the ability to continue long-term passaging.

Note 11: If several large colony forming unit-fibroblasts (CFUs) are observed prior to the cells reaching 70%–80% confluence, the cells can be disassociated on their plate, following a procedure similar to that described here with the following modifications: following disassociation of the cells via trypsin, add the MSC media directly to the plates and gently swirl the dish to separate the cells along the surface, as described when plating the MSCs in this protocol. This can be considered half a passage.

Note 12: Use the least amount of trypsin possible for the shortest amount of time possible to remove the cells from the base of the dish: as a proteolytic enzyme, trypsin can cleave itself. Subsequently, using excess trypsin can be counterproductive. Leaving the trypsin on for too long can also result in excess damage to the MSCs, shortening their viability as stem cells in culture. The dishes must be incubated at 37 °C to permit the trypsin enzyme to work.

Note 13: A small amount of serum-containing media is added to the tubes prior to centrifugation to inactivate any residual trypsin in the solution.

Note 14: When counting MSCs, manual counting with a hemocytometer is preferred versus the use of an automated cell counter: because MSCs are a heterogeneous population encompassing cells of various sizes, readings with automated counters do not give reproducible results. While trypan blue or other cell viability dyes can be used to assess the cell's viability, this is not a critical step with primary MSCs: the viability is generally >99%.

Note 15: Using a combination of previously used MSC media (from the same batch of MSCs) with fresh MSC media will return some of the MSC secretomes from the previous passaged cells. We have found that reuse of media enhances the proliferation and sustained multi-potency of MSCs. Similarly, the dishes can be used several times, returning the MSCs to the extracellular matrix they previously produced.

# References

[1] Dominici M, Le Blanc K, Mueller I, Slaper-Cortenbach I, Marini F, Krause D, Deans R, Keating A, Prockop D and Horwitz E. Minimal criteria for defining multipotent mesenchymal stromal cells. The

International Society for Cellular Therapy position statement. Cytotherapy. 2006; 8(4):315–317.

[2] Sherman LS, Shaker M, Mariotti V and Rameshwar P. Mesenchymal stromal/stem cells in drug therapy: New perspective. Cytotherapy. 2017; 19(1):19–27.

[3] Mohamed-Ahmed S, Fristad I, Lie SA, Suliman S, Mustafa K, Vindenes H and Idris SB. Adipose-derived and bone marrow mesenchymal stem cells: a donor-matched comparison. Stem Cell Res & Ther. 2018; 9(1):168.

[4] Romagano MP, Sherman LS, Shadpoor B, El-Far M, Souayah S, Pamarthi SH, Kra J, Hood-Nehra A, Etchegaray JP, Williams SF and Rameshwar P. Aspirin-Mediated Reset of Preeclamptic Placental Stem Cell Transcriptome - Implication for Stabilized Placental Function. Stem Cell Rev Rep. 2022; 18(8):3066–3082.

[5] Fernandez-Moure JS, Corradetti B, Chan P, Van Eps JL, Janecek T, Rameshwar P, Weiner BK and Tasciotti E. Enhanced osteogenic potential of mesenchymal stem cells from cortical bone: a comparative analysis. Stem Cell Res & Ther. 2015; 6(1):203.

# 4

# Manual Isolation and Expansion of Adipose-derived Stem Cells from Adipose Tissue

**Lauren S. Sherman[1,2], Derek J. Woloszyn[3], Pranela Rameshwar[1], and Edward S. Lee[3]**

[1]Department of Medicine, Rutgers New Jersey Medical School, USA
[2]Rutgers School of Graduate Studies at New Jersey Medical School, USA
[3]Department of Surgery, Rutgers New Jersey Medical School, USA
**Corresponding Author:** Edward S. Lee, Department of Surgery, Rutgers New Jersey Medical School, USA

## Abstract

Adipose-derived stem cells (ASC), a class of adult mesenchymal stem cells, hold great promise for biomedical sciences – both for basic science research and for clinical application. As surgically discarded tissue from a multitude of elective procedure, which frequently involves young otherwise healthy patients, adipose tissue is a valuable, easy to obtain source from which these stem cells can be obtained. This technique describes a method for isolating and expanding human ASCs from solid fat or from lipoaspirate.

## 4.1 Introduction

Adipose-derived stem cells (ASCs) are an attractive source for regenerative, reconstructive, and oncologic medicine. Among the factors making ASCs so attractive is their ease of isolation from readily available and harvested adipose tissue, either lipoaspirate or bulk tissue, and their capacity to be licensed, or educated, into an anti-inflammatory cell type following exposure to an inflammatory microenvironment. Various methods can be used to isolate the ASCs from either of these tissue types, either within the operating theater by centrifugation or in a laboratory using enzyme or enzyme-free methods. In each case, the ASCs are isolated from the stromal vascular fraction (SVF)

29

of the adipose tissue; after culturing the SVF as descried below, the ASCs remain while contaminating cell types (e.g., adipocytes, erythrocytes, endothelial cells, and connective tissues) are removed.

Like other sources of mesenchymal stem cells (MSCs), ASCs are adult stem cells capable of self-renewal, are plastic adherent, can hone to areas of inflammation, show multilineage differentiation (at a minimum into adipocytes, chondrocytes, and osteocytes), and express cell surface markers including CD34⁻, CD45⁻, HLA-DR⁻, CD44⁺, CD73⁺, CD90⁺, CD105⁺, STRO-1⁺[1, 2].

While enzyme-based methods initially produce higher volumes of ASCs as compared to enzyme-free methods, the stress of the enzymatic digestion causes the ASCs not to passage as long as their (non-?)enzymatically digested counterparts [3] However, at times, an initial large-scale ASC isolation is preferable, as in cases where the adipose is removed and ASCs are isolated for immediate autologous injection.

## 4.2 Materials

*All materials should be sterile and used under aseptic conditions unless stated otherwise.*

1.   Approval from your institute's Institutional Review Board (IRB) or equivalent.

2.   Storage bag or container to transport adipose from the operating room to the laboratory (the lipoaspirate suction container works well for this).

3.   Sterile saline (any normal saline is acceptable, including phosphate buffered saline).

4.   Dulbecco's Modified Eagle Medium, with high glucose and sodium bicarbonate, without L-glutamine and sodium pyruvate (DMEM; Sigma Aldrich D5671 or equivalent).

5.   L-glutamine.

6.   Penicillin–streptomycin antibiotic solution.

7.   Fetal bovine serum (Note 1).

8.   100 mm round, plasma-gas-treated tissue culture dishes (Falcon 353003 or equivalent, Note 2).

9.   Phosphate buffered saline (PBS).

10. 0.05% Trypsin-EDTA solution (trypsin).

11. 15 and/or 50 ml centrifuge tubes.

12. Large-bore pipette (10 or 25 ml).

13. For lipoaspirate, optional:

    a. Syringe or beaker (size based on tissue volume).

    b. 16-guage needle.

14. For bulk adipose tissue:

    a. Scissor.

    b. Scalpel.

    c. Forceps and/or a needle.

    d. Cutting surface (e.g., Petri dish lid).

15. For enzymatic digestion:

    a. Collagenase Type-2, 0.075% (Note 3)

    b. 50 ml tubes

    c. Biohazard bag

    d. 200 μm filter

    e. Cotton swab (optional)

## 4.3 Methods

### 4.3.1 Preparation

1. Obtain informed consent if required by your IRB.

2. Transfer lipoaspirate from the operating room to the laboratory.

    i. If the sample will not be processed immediately, it is preferable to store the sample moist: normal saline can be added to the adipose container to keep bulk tissue wet.

    ii. Samples should be stored at room temperature if they will be processed that day. Alternatively, the sample may be stored at 4 °C for up to 72 hours.

3.  Prepare complete media containing DMEM supplemented with 10% FBS, 1% L-glutamine, and 1% penicillin–streptomycin solution

4.  Prepare tissue culture plates, adding approximately 10 ml tissue culture media to each plate

## 4.3.2 Lipoaspirate preparation

1.  Depending on the aspiration vessel used to collect the lipoaspirate, it may be preferable to transfer the lipoaspirate to a syringe or sterile beaker prior to continuing. If transferring to a syringe, a 16-guage needle or other large bore needle can be used.

2.  Allow sample to separate into three layers: an upper layer of oil; a middle layer containing adipose tissue; and a lower layer containing saline and other contaminating cells, including red blood cell. The layers may differ by volume based on the harvesting technique.

3.  Separate the adipose layer from the oil and saline layers. The method of separation will differ based on the aspirating vessel used to harvest the lipoaspirate:

    A.  Syringe or other vessel, which is accessed from the bottom:

        i.  Expel majority of the lower saline layer into a waste receptacle.

        ii.  Slowly expel the adipose into the tissue culture plates, adding approximately 5 ml tissue per plate.

        iii.  Attempt to expel the smallest amount of saline solution into the tissue culture dishes.

    B.  Beaker or other vessel accessed from the top:

        i.  Carefully aspirate the oil layer from above the adipose. Then either:

            1.  Using a large pipette (10–25 ml) remove the adipose layer and transfer approximately 5 ml to each tissue culture plate.

            2.  Place the aspirator tip at the bottom of the saline layer and carefully aspirate the saline layer. Then transfer the adipose to the tissue culture dish as in (A) or by carefully decanting.

4.  Continue with Section 4.3.4 or 4.3.5.

### 4.3.3 Block adipose preparation

1. Cut a manageable segment of tissue from the tissue block and transfer it onto the cutting surface (i.e., Petri dish lid).

2. Using the forceps or needle to hold the tissue in place, use the scalpel to gently scrape open the adipose lobules and separate from the connective tissue scaffold (Note 4). If utilizing enzymatic digestion, the tissue may simply be minced into pieces of approximately 2 mm², taking care to avoid as much fibrous tissue as possible.

3. Once a sufficient volume of adipocytes have been released, transfer the adipocytes to a 50 ml tube to contain the tissue until you are done processing the adipose block.

4. Continue with Section 4.3.4 or 4.3.5.

### 4.3.4 Enzymatic digestion of adipose tissue (optional)

*Optionally, the prepared lipoaspirate or block adipose tissue can be enzymatically digested to enhance the rapid isolation of ASCs.*

1. Transfer the prepared adipose from Section 4.3.2 or 4.3.3 to 50 ml tubes, adding approximately 25 ml of tissue.

2. Add 25 ml of the collagenase solution to each tube (final ratio 1:1).

3. Incubate the tubes at 37 °C for 45 minutes, shaking the tubes every 5–10 minutes (Notes 5–7).

4. Filter the digested sample through the 200 µm filter into fresh 50 ml collection tubes. If the filter becomes clogged, gently clear it with a pipette tip or cotton swab; if the filter remains clogged, use a fresh filter (Note 8).

5. Centrifuge the tubes at 300× *g* for 5–10 minutes.

6. Aspirate the supernatant and resuspend the cell pellet in PBS to wash out the collagenase.

7. Centrifuge the tubes again at 300× *g* for 5–10 minutes.

8. Resuspend the cell pellet with complete media in the original volume (i.e., resuspend 25 ml of tissue in 25 ml complete media).

### 4.3.5 ASC culture and expansion

1. Transfer 5 ml of the adipocyte solution to each tissue culture plate and transfer the plated ASCs to a 37 °C, 5% $CO_2$ incubator (Note 9).

2. After three days, remove a small amount of media from each plate and add a similar amount of fresh complete media.

3. Monitor the cells every 2–3 days to assess if ASCs are visible on the plates.

4. Once cells are visible or at 7–10 days post-ASC isolation, whichever is earlier, remove the tissue using a large bore pipette (10 or 25 ml) and add fresh complete media.

5. See Section 4.3.4 "Culture of MSCs" for the remainder of the culture and expansion protocol.

6. After the third passage, the ASCs can be phenotyped and functionally confirmed by multilineage differentiation as described in Chapters ___.

7. ASCs can be cryopreserved as per the MSC method described in Chapters ___.

## 4.4 Notes

Note 1: FBS must be batch-tested to ensure it can support the growth of ADSCs; in our experience, FBS that supports MSCs will support ADSCs as well.

Note 2: In our hands, the MSCs grow with better efficiency and longevity in tissue culture dishes as compared to tissue culture flasks.

Note 3: Collagenase solutions may be stored at 4 °C for short-term use or −80 °C for long-term storage. Avoid repetitive freeze-thaw cycles so as not to lose the enzymatic activity.

Note 4: Video demonstrations of shearing the lobules are available online [4].

Note 5: The 37 °C incubation can be done in an incubator or in a water bath. It is recommended to place the tubes into a biohazard bag to mitigate chances of contamination of the sample by the water bath or of the water bath or incubator by the tissue sample.

Note 6: The tubes are shaken every 5–10 minutes to ensure the tissue maintains contact with the collagenase. The tubes should be manually shaken every several minutes regardless of whether they are being mechanically agitated by the water bath or incubator during this time. This is of particular importance with block adipose samples.

Note 7: If the sample does not appear to be well digested after 45 minutes, it can be incubated for an additional 15 minutes. After 1 hour, if large pieces of tissue are still present, they should either be collected in Step 3.4.4 and digested in fresh collagenase or discarded. If the pieces of tissue are particularly large, they should be minced into smaller pieces as in Section 4.3.3.

Note 8: If several large pieces of tissue remain, optionally, the filtering step can be skipped, moving directly on to the centrifugation step.

Note 9: Do not stack dishes in columns of greater than 4: if the column is too large, the lower plates will not receive adequate airflow for gas exchange.

## References

[1] Dominici, M., et al., *Minimal criteria for defining multipotent mesenchymal stromal cells. The International Society for Cellular Therapy position statement.* Cytotherapy, 2006. 8(4): p. 315–7.

[2] Sherman, L.S., et al., *Mesenchymal stromal/stem cells in drug therapy: New perspective.* Cytotherapy, 2017. 19(1): p. 19–27.

[3] Sherman, L.S., et al., *Enzyme-Free Isolation of Adipose-Derived Mesenchymal Stem Cells.* Methods Mol Biol, 2018. 1842: p. 203–206.

[4] Sherman, L.S., et al., *An Enzyme-free Method for Isolation and Expansion of Human Adipose-derived Mesenchymal Stem Cells.* J Vis Exp, 2019(154).

# 5

# Isolating Compact Bone-derived Mesenchymal Stem Cells from Rodent and Rabbit Femurs

**Najerie McMillian, Joseph S. Fernandez-Moure, and Taylor Hudson**

Division of Trauma, Acute, and Critical Care Surgery, Department of Surgery, Duke University School of Medicine, USA
**Corresponding Author:** Joseph S. Fernandez-Moure, Department of Surgery-Trauma, Acute and Critical Care Surgery, Duke University Medical Center, USA
E-mail: joseph.fernandezmoure@duke.edu
**Disclaimer:** The authors declare no conflict of interest.

## Abstract

Mesenchymal stem cells (MSCs) are multipotent stems cells that self-renew and differentiate into specialized cells such as chondrocytes (cartilage cells), osteoblasts (bone cells), and adipocytes (fat cells). MSCs can be isolated from various tissues throughout the body including the periosteum, muscle, circulating blood, bone marrow, adipose tissue, blood vessels, and trabecular bone. Recently, scientists have successfully extracted and cultured MSCs from compact bones. There are various methods for isolating compact bone (CB) MSCs from human, rabbit, and rodent models and have been successfully utilized in clinical applications. This protocol describes isolating CB-MSCs from rabbit and rat animal models.

## 5.1 Introduction

Mesenchymal stem cells (MSCs) have demonstrated promising regenerative abilities, making them an excellent candidate for tissue repair studies in medicine, which includes cell-based therapies. The attractiveness of MSCs is due to their pluripotency, lack of histocompatibility complexes, immune stimulating molecules, and immunosuppressive and anti-inflammatory capabilities [1, 2]. Enhanced osteogenic potential of MSCs from cortical bone using a comparative analysis indicated that these cells can be extracted from almost all types of tissues, but the most common source of MSCs are bone marrow (BM) and adipose tissue [3].

Overwhelming studies with MSCs discovered their potential to treat different diseases. These promising effects lie in their ability to self-renew, differentiate into several cell lineages, and participate in immunomodulation [4]. These cells can rapidly multiply in a process of self-regulation, travel to the injury site, and assist in damaged tissue repair without triggering a harmful immune response from the host immune system, due to their lack of histocompatibility complexes. While these abilities give MSCs much translational potential, more research is necessary to improve current limitations. For example, stem cell's ability to differentiate and proliferate may vary considerably based on the originating tissue of the MSCs [4]. This is thought to be due to their primary microenvironment, which influences cell differentiation and signaling to the cell, in which these functions would be beneficial to the native tissue. Thus, tissue specificity is critical when considering the use of MSC in regenerative medicine, and the specific applications.

BM is a common source of MSCs used in clinical applications, though some studies reported that MSCs isolated from compact bone fragments are more beneficial than BM-MSCs [5]. Similar to BM-MSCs, MSCs from compact bone (UC-MSCs) and adipose tissue are immunosuppressive, multipotent with excellent proliferative abilities [5]. Additionally, MSCs exhibit excellent immunomodulatory properties, have slower proliferation rates at earlier passages, and have greater commitment toward the osteogenic lineage [5]. There is even evidence that CB-MSCs have better regenerative properties than BM-MSCs.

We studied the osteogenic potential of MSCs isolated from cortical bone fragments compared to BM-MSCs or adipose-derived MSCs [2]. The study found that, compared to BM-MSCs and adipose-derived MSCs, cells isolated from cortical bone exhibited higher alkaline phosphatase (ALP) activity, which is a biochemical marker for osteoblast activity, mineral disposition, and osteogenic gene expression. After osteogenesis induction, CB-MSCs had greater calcium deposition at both two and four weeks. *ALP* and *RUNx2* are genes that regulate osteoblastogenesis, while *SPP1* and *Bglap* are associated with differentiation. These genes were upregulated in CB-MSCs. Furthermore, hypoxic studies revealed that these cells exhibited greater survival in an oxygen and nutrient-deprived environment, which is beneficial for successful tissue repair and angiogenesis [2]. These results indicate that CB-MSCs could be a promising candidate for supplemental traumatic injury treatment, critical size defects, and orthopedic injuries [2].

To characterize MSCs from compact bones, they must meet the following criteria: adherence to plastic surface of tissue culture flask, CD105, CD73, and CD90 expression while lacking CD45, CD34, CD14, CD11b, CD79a, or CD19 and HLA class II expression, and ability to differentiate osteoblasts, adipocytes, and chondroblasts [6]. CB-MSCs that have been isolated from rat femurs have thus far adhered to the flask surface, have been induced to differentiate osteoblasts and adipocytes, and expressed CD90 and CD73 and not CD45, CD34, CD11b, CD19, and HLA class II.

The isolation of CB-MSC has been shown in both rats and human tissues, paving the way for clinical translation. This evidence suggests that cortical bone is an excellent source of MSCs and can be isolated with minimal invasiveness intraoperatively, i.e., fracture fixation, laminectomy, etc. [3]. These methods yielded more abundant MSCs with greater osteogenic potential and activity than cells obtained from other tissue types [7]. This evidence seems to suggest these techniques could further enhance future research in this field. Below we described our adapted protocol for the isolation of CB-MSC.

## 5.2 Materials

### 5.2.1 Reagents for CBM-MSC isolation

i. Minimal essential medium (MEM) α, L-glutamine, ribonucleosides and deoxyribonucleosides, and 0.1 μm sterile filtered (Cytiva, Hyclone Laboratories, Logan, Utah, 84321).

ii. Premium grade fetal bovine serum (FBS) (VWR International, catalogue# 114-057-131).

iii. 0.25% Trypsin-EDTA (1X) (Thermofisher, catalogue# 25200-056).

iv. Penicillin–streptomycin Gibco Amphotericin B (Thermofisher, catalogue# 15240-062).

v. Collagenase type I or type II.

vi. Dulbecco's phosphate buffered saline (Quality Biological, Gaithersburg, MD, 20879).

vii. 70% Ethanol.

## 5.2.2 Animals

The type of animal model will depend on your study. However, this method can be adapted to any bone. The technique below describes the method to isolate CB-MSCs from rat femur.

## 5.2.3 Materials

   i.  6-well plates
  ii.  Petri dishes
 iii.  50 ml polypropylene tubes
  iv.  Syringes
   v.  23 g needles
  vi.  Microdissecting scissors or bone shears
 vii.  Forceps
viii.  Sterile gauze
  ix.  Incubator
   x.  Biosafety cabinet

# 5.3 Methods

## 5.3.1 Isolating CBM MSCs

1.  Terminate the animal with an IACUC approved method; this will depend on the animal.

2.  Thoroughly rinse the designated leg with 70% ethanol. Shaving the leg is optional.

3.  Carefully dissect the femur at the hip and knee joints using a surgical blade and forceps. Carefully pull back the skin and remove the muscle and all soft tissues. Take care to maintain sterility as contamination when handling live tissue can occur.

4.  After dissecting, free the bones, carefully wipe them with sterile gauze, wash with DPBS, and store them in α-MEM media supplemented with 1% antibiotic/antimycotic and fetal bovine serum. Be sure to store the tube on ice. This will preserve the cells up to 2 hours. Extract femurs within 30 minutes of animal death to ensure high cell viability.

5. In a biosafety cabinet, wash the bones twice with PBS to flush away any remaining blood cells and soft tissues.

6. Using forceps, carefully hold the bone and cut both ends below the end of the marrow cavity using bone shears or microdissecting scissors. This will depend on the bone size.

7. Transfer the bones to a 100 mm culture dish with 10 ml α-MEM media.

8. Using a 23-gauge needle attached to a 5 ml syringe, draw up 5 ml of complete α-MEM medium and slowly flush out the bone cavities. Flush the bones repeatedly until they are pale, ensuring complete mechanical removal of all bone marrow cells and tissue. Discard the aspirate.

9. For rat femurs, crush the bones into 1–3 mm² chips. Typically, other bone fragments can be crushed into chips approximately 3–4 mm³.

10. Suspend bone chips in α-MEM containing 2% (v/v) defined fetal bovine serum (FBS) in the presence of 3 mg/mL collagenase type I and 4 mg/mL dispase II and then placed on a shaking platform at 37 °C for 3 hours.

11. Following digestion, bone chips were plated into new flasks and incubated at 37 °C in a 5% $CO_2$ incubator. Incubate in α-MEM containing 2% (v/v) FBS undisturbed for three days to allow cells to migrate out of the fragments and adhere to the plate.

12. Histology can be performed on digested bone pieces to assess efficacy of digestion and isolation.

13. After 72 hours, remove non-adherent cells by washing the plate with PBS and add fresh complete medium. Change medium every 48 hours as needed.

14. After reaching 70%–90% confluence, cells can be released with trypsin and plated into bigger flasks, such as a T-125.

   a. After removing the medium and washing the cells, digest cells with 0.5 ml of 0.25% trypsin for about 2 minutes. Do not incubate the cells with trypsin for a long time.

   b. Add 1 ml of complete medium to neutralize the trypsin. Split the cells and plate them into a bigger flask.

   c. After reaching 70%–90% confluence, split cells at 1:3 or a desired ratio. Count and cryopreserve the remaining cells.

## 5.4 Notes

Note 1: Before performing any animal experiments, make sure you have received approval from your Institutional Animal Care and Use Committee (IACUC) and always follow all experimental regulations and rules.

Note 2: Always minimize animal suffering. When euthanizing animals, use IACUC-approved methods, as outlined in your approved protocol.

Note 3: Huang et al. describe common challenges and possible solutions in BM-MSC culture. Consult this article when experience issues [7].

Note 4: MSCs are very sensitive to trypsin. Do not incubate cells in trypsin for long periods of time.

Note 5: Cells can be passaged up to four times before losing stemness. Use discretion to determine if you should end the passaging and to start a new culture when splitting the cells.

Note 6: Use fresh cells when characterizing cells.

Note 7: Monitor cells for signs of contamination.

## References

[1] Han Y, Li X, Zhang Y, Han Y, Chang F and Ding J. Mesenchymal stem cells for regenerative medicine. Cells. 2019; 8(8):886.
[2] Fernandez-Moure JS, Corradetti B, Chan P, Van Eps JL, Janecek T, Rameshwar P, Weiner BK and Tasciotti E. Enhanced osteogenic potential of mesenchymal stem cells from cortical bone: a comparative analysis. Stem Cell Res & Ther. 2015; 6(1):1–13.
[3] Fernandez-Moure JS, Corradetti B, Janecek T, Eps JV, Burn M, Weine BK, Rameshwar P and Tasciotti E. Characterization of mesenchymal stem cells from human cortical bone. Intl J Transl Sci. 2016; 2016(1):71–86.
[4] Klimczak A and Kozlowska U. Mesenchymal stromal cells and tissue-specific progenitor cells: their role in tissue homeostasis. Stem Cells Intl. 2016; 2016.
[5] Anastasio A, Gergues M, Lebhar MS, Rameshwar P and Fernandez-Moure J. Isolation and characterization of mesenchymal stem cells in orthopaedics and the emergence of compact bone mesenchymal stem cells as a promising surgical adjunct. World J Stem Cells. 2020; 12(11):1341.

[6] Dominici M, Le Blanc K, Mueller I, Slaper-Cortenbach I, Marini F, Krause D, Deans R, Keating A, Prockop D and Horwitz E. Minimal criteria for defining multipotent mesenchymal stromal cells. The International Society for Cellular Therapy position statement. Cytotherapy. 2006; 8(4):315–317.

[7] Zhu H, Guo Z-K, Jiang X-X, Li H, Wang X-Y, Yao H-Y, Zhang Y and Mao N. A protocol for isolation and culture of mesenchymal stem cells from mouse compact bone. Nature Protocols. 2010; 5(3):550–560.

# 6

# Isolating Dental Pulp Stem Cells

**Ioanna Tsolaki[1,2], Darling Rojas[3], Adam Eljarrah[4], and Pranela Rameshwar[2,4]**

[1]Department of Periodontics, Rutgers School of Dental Medicine, USA
[2]Rutgers School of Graduate Studies at New Jersey Medical School, USA
[3]Rutgers School of Dental Medicine, USA
[4]Department of Medicine, New Jersey Medical School, USA

**Grant:** The work was supported by a training R25 award from the National Heart Lung and Blood Institute.

**Corresponding Author:** Ioanna Tsolaki, Rutgers School of Dental Medicine, USA
E-mail: ioanna.tsolaki@rutgers.edu

**Disclaimer:** The authors have no conflict of interest.

## Abstract

Dental pulp stem cells (DPSCs) derive from cranial neural crest cell populations. They satisfy the minimum criteria that define mesenchymal stem cells (MSCs). At this time, DPSCs do not seem to express a marker that exclusively identifies them. DPSCs can differentiate into multiple cell types such as neural ectodermal cells, adipocytes, odontoblasts, osteoblasts, chondrocytes, myoblast cells of mesodermal origin, hepatocytes, and endothelial cells. The relatively easy and minimally invasive process of retrieval and isolation, the similarities to the human bone-marrow-derived MSCs, and the multipotent differentiation capabilities of DPSCs make them an attractive stem cell population for research and therapeutic use. The purpose of this chapter is to describe the method to isolate and culture DPSCs from extracted teeth.

## 6.1 Introduction

The dental pulp contains various types of cells, fibrous matrix, and ground substance. The dental pulp cells are dental pulp stem cells (DPSCs), fibroblasts, endothelial cells, odontoblasts, Schwann cells, immune cells, epithelial-like cells, and erythrocytes. DPSCs derive from cranial neural crest cell populations [1]. Mesenchymal stem cells (MSCs) were isolated from the dental pulp of human third molars [2].

MSCs are heterogeneous multipotent cells that can be isolated from various tissues including dental pulp, bone marrow, adipose tissue, placenta, and umbilical cord [2, 3]. The minimum criteria for defining MSCs include cellular adherence to plastic in standard culture conditions, expression of cell surface markers CD105, CD73, and CD90, negative for CD45, CD34, CD14, CD11b, CD79-alpha, or CD19, and low for HLA-DR, and multilineage differentiation into osteoblasts, adipocytes, and chondroblasts [4]. DPSCs meet these minimal criteria [5]. They can differentiate into neural ectodermal cells, adipocytes, odontoblasts, osteoblasts, chondrocytes, myoblast cells of mesodermal origin, hepatocytes, and endothelial cells [2, 6–13].

Similar to MSCs from other tissues, DPSCs display low immunogenicity and exhibit immunomodulatory properties via cell-contact-dependent mechanisms and through the secretion of soluble factors and extracellular vesicles [14–18]. These properties make DPSCs an attractive tool in tissue engineering and regenerative medicine, including in the context of oral-maxillo-facial bone repair and dental pulp regeneration [7, 19]. Additionally, MSCs play key roles in the maintenance of both hematopoietic stem cells (HSCs) and leukemic stem cells and support the transition of differentiated cancer cells into dormancy [20, 21]. The presence of HSCs in dental pulp could then suggest a role for DPSCs in stem cell maintenance during both healthy and diseased states [22].

At this time, DPSCs do not seem to express a marker that exclusively identifies them from other areas within the gingiva. The identification of DPSCs are similar to MSCs and are therefore characterized phenotypically based on the expression of CD73, CD90, and CD105, as well as other markers used in studies with MSCs [10]. A few studies performed single-cell RNA-sequencing of the dental pulp [23–25]. Single-cell RNA sequencing analysis of the dental pulp from five human third molars indicated higher expression of FRZB, NOTCH3, THY1, and MYH11 as compared to other dental pulp cells [23].

DPMSCs are an important source of MSCs due to the relatively easy and minimally invasive process of retrieval and isolation, their similarities to the human bone-marrow-derived MSCs, and their multipotent differentiation capabilities [4, 9–13, 26–28]. Regarding DPMSC location within the dental pulp, it was initially suggested that DPMSCs originate from niches within the cell-rich sub-odontoblast layer, the dental pulp stroma, and especially from perivascular regions surrounding the pulpal vasculature. The current theory is that there are several stem cells niches within the dental pulp [29–37]. This study specifically describes the method to isolate and culture DPSCs from extracted teeth.

## 6.2 Materials

### 6.2.1 Extracted human teeth with vital pulp (Note 1)
### 6.2.2 Dental instruments

   i. Chisel for splitting teeth

  ii. Mallet

 iii. Periodontal probe UNC15

 iv. High-speed handpiece

  v. Carbide bur 557 FG

 vi. Micro Adson tissue pliers

### 6.2.3 DPSCs culture media

   i. 90% DMEM

  ii. 10% FBS

 iii. 1% L-glutamine

 iv. 1% AA

  v. Fungizone (0.25 µg/ml)

### 6.2.4 Flow cytometry for phenotype

CD73(+), CD90(+), CD105(+), and CD45(−) (Refer to Method 16 for a detailed protocol).

## 6.3 Methods

### 6.3.1 Tooth preparation (Figure 6.1)

   i. Extract the tooth.

  ii. Rinse the extracted tooth with sterile tissue culture grade 1× PBS (Notes 2 and 3).

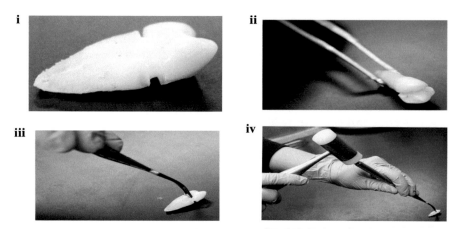

**Figure 6.1**   Tooth sectioning. Grooves are prepared along the cemento-enamel junction (i), the long axis of each root (i), and the mesial, occlusal, and distal surface of the crown of the tooth (ii). The tooth is sectioned along the prepared grooves using a bone chisel (iii, iv) and a dental mallet (iv).

iii.   Drill grooves with a high-speed dental hand-piece and sterile carbide bur along the buccal and lingual/palatal surface of each of the roots and the crown of the tooth. Additionally, prepare a groove at the perimeter of the anatomic crown of the tooth at the level of the cemento-enamel junction. The depth of the grooves is in dentin.

iv.   Prepare a 6-well plate with one cavity with pure Povidone-iodine, four cavities with sterile PBS, and one cavity with A/A (100×) and Fungizone (0.25 µg/ml).

v.   Incubate the tooth in pure Povidone-iodine for 4 minutes.

vi.   Wash the tooth with sterile PBS until no brown residue leftover is seen.

vii.   Incubated the tooth in A/A (100×) for 4 minutes.

viii.   Wash the tooth with sterile PBS in the residual 3 cavities with PBS.

### 6.3.2 Dental pulp extraction

i.   Fracture the tooth along the prepared grooves. Use chisel and dental mallet for this purpose. Use controlled force in order to avoid damaging the dental pulp.

ii.   Extract the dental pulp using Micro Adson tissue pliers and periodontal probe NC15.

### 6.3.3 Dental pulp preparation

i. Place the extracted dental pulp in an empty tissue culture plate.

ii. Cut the extracted dental pulp in a few pieces.

iii. Let the pieces of dental pulp dry for at least 10 minutes.

iv. Add the DPSCs media until the pieces of dental pulp tissue are just covered.

v. Replace the media with 50% fresh media at 3–4 days' intervals.

vi. The DPSCs appeared will show spindle shaped cells at approximately seven days.

vii. Split the DPSCs cells at 75%–80% of confluence (Notes 4–6).

## 6.4 Notes

Note 1: The age, dental pulp health, and genetic background of the donor can affect the amount of dental stem cells isolated.

Note 2: The extracted teeth are transferred in PBS immediately after removal from the patient.

Note 3: The extracted teeth can be stored at 4 °C (refrigerator or cold room). We have found that you can recover a culture of DPSCs even after 2–3 days of storage.

Note 4: Anti-fungal drug is optional after the second passage of the isolated dental pulp stem cells.

Note 5: There are no markers specific to dental pulp stem cells.

Note 6: Cryopreservation: Add equal volume of chilled cryopreservation media to DPSCs, resuspended at ~5 × $10^7$ cells/ml. Gently shake the cells while adding the media. Place the cells in a controlled rate freezer (Cryo Met Freezer, Thermo Fisher) at −1 °C/minute until the temperature reaches −100 °C. After this, transfer the cells into liquid nitrogen (refer to Method 17 for a detailed protocol).

## References

[1] Luan X, Dangaria S, Ito Y, Walker CG, Jin T, Schmidt MK, Galang MT and Druzinsky R. Neural crest lineage segregation: a blueprint for periodontal regeneration. J Dent Res. 2009; 88(9):781–791.

[2] Gronthos S, Mankani M, Brahim J, Robey PG and Shi S. Postnatal human dental pulp stem cells (DPSCs) in vitro and in vivo. Proc Natl Acad Sci U S A. 2000; 97(25):13625–13630.

[3] Pittenger MF, Discher DE, Péault BM, Phinney DG, Hare JM and Caplan AI. Mesenchymal stem cell perspective: cell biology to clinical progress. NPJ Regen Med. 2019; 4:22.

[4] Dominici M, Le Blanc K, Mueller I, Slaper-Cortenbach I, Marini F, Krause D, Deans R, Keating A, Prockop D and Horwitz E. Minimal criteria for defining multipotent mesenchymal stromal cells. The International Society for Cellular Therapy position statement. Cytotherapy. 2006; 8(4):315–317.

[5] Özdemir AT, Özgül Özdemir RB, Kırmaz C, Sarıboyacı AE, Ünal Halbutoğlları ZS, Özel C and Karaöz E. The paracrine immunomodulatory interactions between the human dental pulp derived mesenchymal stem cells and CD4 T cell subsets. Cell Immunol. 2016; 310: 108–115.

[6] Rodas-Junco BA, Canul-Chan M, Rojas-Herrera RA, De-la-Peña C and Nic-Can GI. Stem Cells from Dental Pulp: What Epigenetics Can Do with Your Tooth. Front Physiol. 2017; 8:999.

[7] Potdar PD. Human dental pulp stem cells: Applications in future regenerative medicine. World J Stem Cells. 2015; 7(5):839.

[8] Longoni A, Utomo L, Van Hooijdonk I, Bittermann G, Vetter V, Kruijt Spanjer E, Ross J, Rosenberg A and Gawlitta D. The chondrogenic differentiation potential of dental pulp stem cells. Eur Cells Materials. 2020; 39:121–135.

[9] Huang GT, Gronthos S and Shi S. Mesenchymal stem cells derived from dental tissues vs. those from other sources: their biology and role in regenerative medicine. J Dent Res. 2009; 88(9):792–806.

[10] Ledesma-Martínez E, Mendoza-Núñez VM and Santiago-Osorio E. Mesenchymal Stem Cells Derived from Dental Pulp: A Review. Stem Cells Intl. 2016; 2016:4709572.

[11] Nuti N, Corallo C, Chan BM, Ferrari M and Gerami-Naini B. Multipotent Differentiation of Human Dental Pulp Stem Cells: a Literature Review. Stem Cell Rev Rep. 2016; 12(5):511–523.

[12] Anitua E, Troya M and Zalduendo M. Progress in the use of dental pulp stem cells in regenerative medicine. Cytotherapy. 2018; 20(4): 479–498.

[13] Shi X, Mao J and Liu Y. Pulp stem cells derived from human permanent and deciduous teeth: Biological characteristics and therapeutic applications. Stem Cells Transl Med. 2020; 9(4):445–464.

[14] Lee S, Zhang QZ, Karabucak B and Le AD. DPSCs from Inflamed Pulp Modulate Macrophage Function via the TNF-α/IDO Axis. J Dent Res. 2016; 95(11):1274–1281.

[15] Wada N, Menicanin D, Shi S, Bartold PM and Gronthos S. Immunomodulatory properties of human periodontal ligament stem cells. J Cell Physiol. 2009; 219(3):667–676.

[16] Li Y, Duan X, Chen Y, Liu B and Chen G. Dental stem cell-derived extracellular vesicles as promising therapeutic agents in the treatment of diseases. Int J Oral Sci. 2022; 14(1):2.

[17] Zhao Y, Wang L, Jin Y and Shi S. Fas ligand regulates the immuno-modulatory properties of dental pulp stem cells. J Dent Res. 2012; 91(10):948–954.

[18] Eljarrah A, Gergues M, Pobiarzyn PW, Sandiford OA and Rameshwar P. Therapeutic Potential of Mesenchymal Stem Cells in Immune-Mediated Diseases. Adv Exp Med Biol. 2019; 1201:93–108.

[19] Liang C, Liao L and Tian W. Stem Cell-based Dental Pulp Regeneration: Insights From Signaling Pathways. Stem Cell Rev Rep. 2021; 17(4):1251–1263.

[20] Forte D, García-Fernández M, Sánchez-Aguilera A, Stavropoulou V, Fielding C, Martín-Pérez D, López JA, Costa ASH, Tronci L, Nikitopoulou E, Barber M, Gallipoli P, Marando L, et al. Bone Marrow Mesenchymal Stem Cells Support Acute Myeloid Leukemia Bioenergetics and Enhance Antioxidant Defense and Escape from Chemotherapy. Cell Metab. 2020; 32(5):829–843.e829.

[21] Sandiford OA, Donnelly RJ, El-Far MH, Burgmeyer LM, Sinha G, Pamarthi SH, Sherman LS, Ferrer AI, DeVore DE, Patel SA, Naaldijk Y, Alonso S, Barak P, et al. Mesenchymal Stem Cell-Secreted Extracellular Vesicles Instruct Stepwise Dedifferentiation of Breast Cancer Cells into Dormancy at the Bone Marrow Perivascular Region. Cancer Res. 2021; 81(6):1567–1582.

[22] Osaki J, Yamazaki S, Hikita A and Hoshi K. Hematopoietic progenitor cells specifically induce a unique immune response in dental pulp under conditions of systemic inflammation. Heliyon. 2022; 8(2):e08904.

[23] Pagella P, de Vargas Roditi L, Stadlinger B, Moor AE and Mitsiadis TA. A single-cell atlas of human teeth. iScience. 2021; 24(5):102405.

[24] Sharir A, Marangoni P, Zilionis R, Wan M, Wald T, Hu JK, Kawaguchi K, Castillo-Azofeifa D, Epstein L, Harrington K, Pagella P, Mitsiadis T, Siebel CW, et al. A large pool of actively cycling progenitors orchestrates self-renewal and injury repair of an ectodermal appendage. Nat Cell Biol. 2019; 21(9):1102–1112.

[25] Krivanek J, Soldatov RA, Kastriti ME, Chontorotzea T, Herdina AN, Petersen J, Szarowska B, Landova M, Matejova VK, Holla LI, Kuchler U, Zdrilic IV, Vijaykumar A, et al. Dental cell type atlas reveals stem and differentiated cell types in mouse and human teeth. Nat Commun. 2020; 11(1):4816.

[26] Shi S, Robey PG and Gronthos S. Comparison of human dental pulp and bone marrow stromal stem cells by cDNA microarray analysis. Bone. 2001; 29(6):532–539.

[27] Sloan AJ and Waddington RJ. Dental pulp stem cells: what, where, how? Int J Paediatr Dent. 2009; 19(1):61–70.

[28] Yang M, Zhang H and Gangolli R. Advances of mesenchymal stem cells derived from bone marrow and dental tissue in craniofacial tissue engineering. Curr Stem Cell Res Ther. 2014; 9(3):150–161.

[29] Yu T, Volponi AA, Babb R, An Z and Sharpe PT. Stem Cells in Tooth Development, Growth, Repair, and Regeneration. Curr Top Dev Biol. 2015; 115:187–212.

[30] Téclès O, Laurent P, Zygouritsas S, Burger AS, Camps J, Dejou J and About I. Activation of human dental pulp progenitor/stem cells in response to odontoblast injury. Arch Oral Biol. 2005; 50(2):103–108.

[31] Fitzgerald M, Chiego DJ, Jr. and Heys DR. Autoradiographic analysis of odontoblast replacement following pulp exposure in primate teeth. Arch Oral Biol. 1990; 35(9):707–715.

[32] Ruch JV. Odontoblast commitment and differentiation. Biochem Cell Biol. 1998; 76(6):923–938.

[33] Smith AJ and Lesot H. Induction and regulation of crown dentinogenesis: embryonic events as a template for dental tissue repair? Crit Rev Oral Biol Med. 2001; 12(5):425–437.

[34] Løvschall H, Tummers M, Thesleff I, Füchtbauer EM and Poulsen K. Activation of the Notch signaling pathway in response to pulp capping of rat molars. Eur J Oral Sci. 2005; 113(4):312–317.

[35] Goldberg M and Smith AJ. Cells and extracellular matrices of dentin and pulp: A Biological Basis for Repair and Tissue Engineering. Crit Rev Oral Biol Med. 2004; 15(1):13–27.

[36] Shi S and Gronthos S. Perivascular niche of postnatal mesenchymal stem cells in human bone marrow and dental pulp. J Bone Miner Res. 2003; 18(4):696–704.

[37] Feng J, Mantesso A, De Bari C, Nishiyama A and Sharpe PT. Dual origin of mesenchymal stem cells contributing to organ growth and repair. Proc Natl Acad Sci U S A. 2011; 108(16):6503–6508.

# 7

# A Murine System for Somatic Cellular Reprogramming into Induced Pluripotent Stem Cells

**Dahui Wang, Edward A. Gonzalez, and Jean-Pierre Etchegaray**

Department of Biological Sciences, Rutgers University, USA
**Grant:** This work was supported by the Bush Biomedical Foundation.
**Corresponding Author:** Jean-Pierre Etchegaray, Rutgers University, Life Sciences Center, USA
E-mail: jeanpierre.etchegaray@rutgers.edu
**Disclaimer:** The authors have nothing to declare.

## Abstract

Somatic cellular reprogramming into induced pluripotent stem cells (iPSCs) is a powerful research tool to investigate the molecular mechanisms underlying cell fate changes. These mechanisms include epigenetic dynamics driven by the pioneer transcription factors Oct4, Sox2, Klf4, and Myc to reprogram overall gene expression profiles. Furthermore, the generation of iPSCs has been instrumental in substantiating the concept of epigenetic memory for the maintenance of cell fate identity, which can change in the context of health and disease. Loss of cell fate identity has been proposed to play a critical role in tumor initiating and propagating programs, possibly due to epigenetic reprogramming pathways that are reminiscent of somatic cellular reprogramming into iPSCs. This level of epigenetic plasticity involved in cell fate changes could facilitate tissue maintenance and repair during development and aging, but also during cellular de-differentiation promoting tumorigenesis. Because of its similarities to embryonic stem cells (ESCs), iPSCs can differentiate into all cell types and thereby hold great potential for regenerative medicine.

## 7.1 Introduction

Under specific conditions, somatic cells can be reprogrammed into pluripotent stem cells capable of generating all cell types that constitute an entire organism. More specifically, somatic cells subjected to forced expression of the pioneer transcription factors Oct4, Sox2, Klf4, and c-Myc, also known as the Yamanaka factors, become induced pluripotent stem cells (iPSCs) in approximately two weeks [1, 2]. iPSCs can be produced by reprogramming various somatic cells including fibroblasts, neural progenitor cells (NPCs), and B cells [3–5]. This chapter describes the process of producing iPSCs from mouse embryonic fibroblasts (MEFs).

Mechanistically, it has been proposed that the epigenome of somatic cells can be reprogrammed to promote two waves of transcriptional modifications that culminate in the formation of iPSCs [6, 7]. During reprogramming, somatic cells go through a gradual process that results in the formation of intermediate cellular states defined by key gene expression changes before becoming iPSCs. More specifically, somatic cells undergoing reprogramming into iPSCs experience two distinct transcriptional waves [7]. The first transcriptional wave, driven by c-Myc and Klf4, is required for erasing somatic cell gene expression, while the second wave, driven by Oct4, Sox2, and Klf4, is responsible for the establishment of epigenetic changes to facilitate the reactivation of the pluripotency gene network [7]. Collectively, epigenetics and transcriptional reconfigurations driven by the Yamanaka factors are needed for the formation of iPSCs [4, 6, 8, 9].

Historically, somatic cellular reprogramming relied on individual expression of the Yamanaka factors using viral infection methods that resulted in low efficiency and reproducibility of iPSC formation. These constraints were partially circumvented by using doxycycline-inducible lentiviral constructs expressing the Yamanaka factors [4, 10–13]. However, limitations to this system include silencing of lentiviral transgenes in a high percentage of iPSC lines and heterogeneous expression of the Yamanaka factors due to integration at random genomic *loci*. Consequently, mice generated from these iPSCs produced heterogeneous offspring due to random genomic integration of lentiviral transgenes. To resolve such limitations, a reprogrammable mouse model was developed from ESCs expressing a doxycycline-inducible polycistronic gene cassette encoding the Yamanaka factors integrated into the 3 -untranslated region of the *Col1a1* gene [9, 14]. These ESCs also express a reverse tetracycline-dependent transactivator (M2-rtTA) from the *Rosa26 locus* to allow doxycycline-depended activation of the Yamanaka factors from the polycistronic cassette [9, 15, 16]. Thus, a key advantage of this

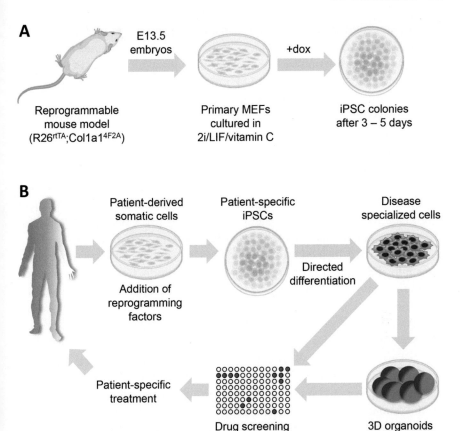

**Figure 7.1**  Generation of iPSCs and its translational applications. (A)Schematic representation of iPSC production from mouse embryonic fibroblasts (MEFs) derived from the reprogrammable mouse model expressing the Yamanaka factors in a doxycycline-inducible manner under 2i/LIF/vitamin C culturing conditions. (B) Schematic diagram showing iPSC application in translational medicine. Generation of patient-specific iPSCs can be subjected to directed differentiation to form patient-specific somatic cells or patient-specific 3D organoids, which can be used for drug screening approaches to develop patient-specific therapies.

reprogrammable mouse model is the induction of iPSCs without viral infection and at a much higher efficiency (Figure 7.1(A)).

Initially, depending on the somatic cell type, generation of iPSCs was accomplished within 14 days at 1%–5% efficiency in regular DMEM [2–4, 7]. The slow reprogramming process was attributed to somatic cells needing to go through a series of intermediate states to eliminate their gene expression pattern and reestablish the pluripotency gene network. Thus, low efficiency of somatic cellular reprogramming toward iPSCs can be a consequence of intermediate cellular stages being incapable of overcoming molecular roadblocks [7]. One of these roadblocks is the persisting expression of genes involved in differentiation such as the fibroblast growth factor 4 (FGF4) [17]. Mechanistically, FGF4 activates MAP kinase signaling cascade composed of mitogen-activated protein kinase kinases (MEK) and extracellular signal-regulated kinases (ERK), which can disrupt ESC renewal [18]. Consequently, a new reprogramming culturing media containing MEK and ERK inhibitors called 2i (two inhibitor) media was developed [17, 18]. In combination with the self-renewal cytokine leukemia inhibitor factor (LIF), the 2i-media induces stable upregulation of the core pluripotency factors Oct4 and Nanog [3]. Under this 2i/LIF conditions, transduction of Oct4 and Klf4 are sufficient to produce iPSCs from neural stem cells (NSCs) [3]. Consequently, the 2i/LIF media, which was originally used to maintain ESC renewal, has been successfully used to increase the efficiency of somatic cellular reprogramming into iPSCs [17]. The frequency of iPSC production can be achieved within four days in 2i/LIF media, as monitored by Oct4-GFP positive colony formation [3, 10].

An additional component that improves the efficiency of somatic cellular reprogramming is vitamin C, and antioxidant and a cofactor for $Fe^{2+}$ and $\alpha$-ketoglutarate ($\alpha$-KG)-dependent dioxygenase enzymes [19, 20]. Mechanistically, vitamin C promotes the efficiency of iPSC formation by enhancing epigenetic reprogramming through activation of the jumonji (JmjC) domain-containing histone demethylases and the ten–eleven translocation (Tet) family of enzymes Tet1, Tet2, and Tet3 [19]. Tet enzymes, initially discovered by Anjana Rao and colleagues, are DNA dioxygenases targeting the sequential oxidation of 5-methylcytosine (5mC) into 5-hydroxymethylcytosine (5hmC), 5-formylcytosine (5fC), and 5-carboxylcytosine (5caC) [21, 22]. Both 5fC and 5caC can be excised through a DNA-based excision repair mechanism that results in demethylated DNA [21]. Earlier, it was found that the replacement of Oct4 by Tet1 could initiate reprogramming into iPSCs that correlated with an enrichment of 5hmC [23]. Mechanistically, it was proposed that Tet1 in combination with the Yamanaka factors are part of a positive feedback loop that promotes DNA demethylation of pluripotent genes [24]. Furthermore, CCAAT/enhancer binding protein-$\alpha$ (C/EBP$\alpha$) was shown to induce the expression and nuclear translocation of Tet2 to promote DNA demethylation on the pluripotency gene network and consequently enhance reprogramming into iPSCs [5, 25]. Consequently, suppression of Tet genes impairs the reprogramming process [26]. Overall, Tet enzymes function synergistically with the Yamanaka factors to facilitate the removal of DNA methylation and prime transcriptional activation of the pluripotency gene network to promote reprogramming into iPSCs [27]. Therefore, 2i/LIF media supplemented with vitamin C enhances iPSC colony formation up to 2%–8.75% within ~4 days [5, 20].

Similar to ESCs, iPSCs can be cultured indefinitely and differentiated into any cell type. Thus, iPSC technology can be applied to patient-personalized disease research. For instance, patient-derived iPSCs can be subjected to cell-directed differentiation for disease modeling and pharmacological screening to establish patient-specific therapies (Figure 7.1(B)). The ultimate goal for using iPSC technology will be patient-personalized cell/tissue transplantation therapy. Here, we describe the protocol for generating iPSCs using MEFs derived from the reprogrammable mouse model mentioned above using the 2i/LIF media supplemented with vitamin C. The advantage for using this mouse model is the availability of diverse somatic cell types such as MEFs, NPCs, and B cells for reprogramming into iPSCs upon doxycycline-mediated induction of the Yamanaka factors (see Notes 1 and 2) [3–5]. Additionally, this model could be used to establish iPSCs from hematopoietic stem cells to study hematopoietic regulation.

## 7.2 Material

### 7.2.1 Reprogrammable mouse model and reagents to prepare murine embryonic fibroblasts (MEFs)

i.   Reprogrammable mouse model $R26^{rtTA}$; $Col1a1^{4F2A}$ (The Jackson Laboratories, Strain No. 011004).

ii.  Dulbecco's phosphate buffer saline (DPBS) without calcium and magnesium (Gibco, Cat. No. 14190144).

iii. Trypsin-EDTA (0.05%) (Gibco, Cat. No. 25300054).

iv.  DMEM (Gibco, Cat. No. 21013-024).

### 7.2.2 Preparation of 2i/LIF media supplemented with vitamin C

i.   Vitamin C (L-ascorbic acid) (Sigma, Cat. No. A92902).

ii.  ERK inhibitor (Sigma, Cat. No. PD0325901).

iii. GSK3 inhibitor (Sigma, Cat. No. CHIR99021).

iv.  Leukemia inhibitory factor (LIF) (Millipore, Cat. No. ESG1107).

   v.  2-mercaptoethanol (Life Technologies, Cat. No. 21985-023).

   vi.  Doxycycline (dox) (Sigma, Cat. No. D9891-100G).

   vii.  KnockOut-DMEM (Gibco, Cat. No. 10829-018).

### 7.2.3 Quantification of iPSC colonies by alkaline phosphatase (AP) staining

   i.  Alkaline phosphatase (AP) staining kit II (Stemgent, 00-0055).

## 7.3 Method

### 7.3.1 Harvesting somatic cells from reprogrammable mouse model (see Notes 1 and 2)

   i.  Collect embryos from the reprogrammable mouse model R26$^{rTA}$; Col1a1$^{4F2A}$ at E13.5.

   ii.  Under sterile conditions, place freshly collected embryos on a culture dish.

   iii.  Cover the embryos with DPBS without calcium–magnesium and remove placenta along with other maternal tissues.

   iv.  Cut away the head and eviscerate the embryos. The head can be used for genomic DNA isolation and genotyping (see Note 3).

   v.  Place the eviscerated embryo's body on a separate culture dish. Each embryo should be handled on an individual culture dish.

   vi.  Add 1 ml of trypsin-EDTA (0.05%) and mince the embryo with sterile scalpel or razor blade.

   vii.  Place culture dish with the minced embryo in a 37 °C incubator for 30–45 minutes.

   viii.  Quench trypsin enzymatic activity by adding 2–4 ml of MEF media [DMEM with 10% fetal bovine serum (FBS), L-glutamine, non-essential amino acids, and penicillin–streptomycin].

   ix.  Place in 37 °C incubator for five days.

   x.  Continue cultures to achieve a uniform monolayer of reprogrammable primary MEFs.

### 7.3.2  Somatic cell reprogramming into iPSCs using 2i/LIF media supplemented with Vitamin C

i. Seed ~$2 \times 10^4$ reprogrammable primary MEFs on each well of a 0.1% gelatin-coated 6-well culture dish containing 2i/LIF/Vitamin C media (KnockOut-DMEM with 15% fetal bovine serum, 1% L-glutamine, 1% non-essential amino acids, 1% penicillin–streptomycin, 0.1 mM 2-mercaptoethanol, 1000 U/ml of LIF, 50 µg/ml of freshly prepared vitamin C, 3 µM of GSK3 inhibitor CHIR-99021, and 1 µM of ERK inhibitor PD0325901).

ii. Culture in 37 °C incubator one day before doxycycline (dox) induction.

iii. Wash cells with PBS and add dox at a final concentration of 1 µg/ml every two days.

iv. Withdraw dox once iPSC colonies appear and continue culturing in 2i/LIF media supplemented with vitamin C.

v. Quantify iPSC colony formation by staining for alkaline phosphatase (AP) activity. Before AP staining, doxycycline is removed for 3–4 days to eliminate exogenous expression of the Yamanaka factors. iPSC efficiency can also be determined by fluorescence (Note 4).

## 7.4  Notes

Note 1: The generation of somatic cells to establish iPSCs can be obtained with other reprogrammable mouse models, which are provided by The Jackson Laboratory:

i. The R26$^{rtTA}$; Col1a1$^{2lox-4F2A}$ (Strain No. 011011) [28]. The Yamanaka factors in the polycistronic transgene cassette of this mouse model are flanked by *loxP* sites and can be excised upon exposure to Cre recombinase. This allows the generation of iPSCs that are free of reprogramming factors.

ii. The iRep1 (inducible reprogrammable mouse 1), also known as Oct4-GFP; ROSA26-rtTA (Δneo); OKMSCh250 (Strain No. 031011). The efficiency of reprogramming and pluripotency status of generated iPSCs can be determined by monitoring the expression of the Yamanaka factors, which is accompanied by mCherry fluorescence as well as monitoring the expression of the Oct4-GFP transgene by EGFP fluorescence.

iii. The Cre-inducible iRep1, also called Oct4-GFP; ROSA26-rtTA (neo); OKMSCh250 (Strain No. 031010). Following Cre recombinase treatment, this mouse model allows the expression of the Yamanaka factors upon doxycycline induction where reprogramming into iPSC can be monitored based on Oct4-GFP-dependent fluorescence.

Note 2: The mouse model described in this protocol can be used for the reprogramming of multiple somatic cell types, e.g., MEFs, neural progenitor cells (NPCs), and B cells. These cells can be reprogrammed into iPSCs according to the procedures described in ref. [3–5].

Note 3: It is convenient to save the head of embryos for genotyping purposes to verify genetic modified genes of the starting reprogrammable tissue material.

Note 4: When using somatic cells derived from iRep1 mouse models (strain Nos 03011 and 031010) expressing fluorescent-tagged Oct4, iPSCs can be further followed by imaging for fluorescein and/or flow cytometry.

## References

[1] Takahashi K, Tanabe K, Ohnuki M, Narita M, Ichisaka T and Tomoda K. et al. Induction of pluripotent stem cells from adult human fibroblasts b y defined factors. Cell. 2007; 131(5):861–872.

[2] Takahashi K and Yamanaka S. Induction of pluripotent stem cells from mouse embryonic and adult fibroblast cultures by defined factors. Cell. 2006; 126 (4): 663–76.

[3] Silva J, Barrandon O, Nichols J, Kawaguchi J, Theunissen TW and Smith A. Promotion of reprogramming to ground state pluripotency by signal inhibition. PLoS Biol. 2008; 6(10):e253.

[4] Stadtfeld M, Maherali N, Breault DT and Hochedlinger K. Defining molecular cornerstones during fibroblast to iPS cell reprogramming in mouse. Cell Stem Cell. 2008; 2(3):230–240.

[5] Sardina JL, Collombet S, Tian TV, Gómez A, Di Stefano B, Berenguer C, Brumbaugh J, Stadhouders R, Segura-Morales C and Gut M. Transcription factors drive Tet2-mediated enhancer demethylation to reprogram cell fate. Cell Stem Cell. 2018; 23(5):727–741. e729.

[6] Hochedlinger K and Plath K. Epigenetic reprogramming and induced pluripotency. Development. 2009: 136(4):509–523.

[7] Polo JM, Anderssen E, Walsh RM, Schwarz BA, Nefzger CM, Lim SM, Borkent M, Apostolou E, Alaei S and Cloutier J. A molecular

roadmap of reprogramming somatic cells into iPS cells. Cell. 2012; 151(7):1617–1632.

[8] Sridharan R, Tchieu J, Mason MJ, Yachechko R, Kuoy E, Horvath S, Zhou Q and Plath K. Role of the murine reprogramming factors in the induction of pluripotency. Cell. 2009; 136(2):364–377.

[9] Stadtfeld M, Maherali N, Borkent M and Hochedlinger K. A reprogrammable mouse strain from gene-targeted embryonic stem cells. Nature Methods. 2010; 7(1):53–55.

[10] Brambrink T, Foreman R, Welstead GG, Lengner CJ, Wernig M, Suh H and Jaenisch R. Sequential expression of pluripotency markers during direct reprogramming of mouse somatic cells. Cell Stem Cell. 2008; 2(2):151–159.

[11] Maherali N, Ahfeldt T, Rigamonti A, Utikal J, Cowan C and Hochedlinger K. A high-efficiency system for the generation and study of human induced pluripotent stem cells. Cell Stem Cell. 2008; 3(3):340–345.

[12] Wernig M, Meissner A, Cassady JP and Jaenisch R. c-Myc is dispensable for direct reprogramming of mouse fibroblasts. Cell Stem Cell. 2008; 2(1):10–12.

[13] Hockemeyer D, Soldner F, Cook EG, Gao Q, Mitalipova M and Jaenisch R. A drug-inducible system for direct reprogramming of human somatic cells to pluripotency. Cell Stem Cell. 2008; 3(3):346–353.

[14] Sommer CA, Stadtfeld M, Murphy GJ, Hochedlinger K, Kotton DN and Mostoslavsky G. Induced pluripotent stem cell generation using a single lentiviral stem cell cassette. Stem Cells. 2009; 27(3):543–549.

[15] Beard C, Hochedlinger K, Plath K, Wutz A and Jaenisch R. Efficient method to generate singleA and Jaenischtton DN and Mostoslavsky G. Induced pluripotent stem cell geneGenesis. 2006; 44(1):23–28.

[16] Hochedlinger K and Jaenisch R. Induced pluripotency and epigenetic reprogramming. Cold Spring Harbor Perspectives in Biol. 2015; 7(12):a019448.

[17] Ying Q-L, Wray J, Nichols J, Batlle-Morera L, Doble B, Woodgett J, Cohen P and Smith A. The ground state of embryonic stem cell self-renewal. Nature. 2008; 453(7194):519–523.

[18] Silva J and Smith A. Capturing pluripotency. Cell. 2008; 132(4):532–536.

[19] Cimmino L, Neel BG and Aifantis I. Vitamin C in stem cell reprogramming and cancer. Trends Cell Biol. 2018; 28(9):698–708.

[20] Esteban MA, Wang T, Qin B, Yang J, Qin D, Cai J, Li W, Weng Z, Chen J and Ni S. Vitamin C enhances the generation of mouse and human induced pluripotent stem cells. Cell Stem Cell. 2010; 6(1):71–79.

[21] Lio C-WJ, Yue X, López-Moyado IF, Tahiliani M, Aravind L and Rao A. TET methylcytosine oxidases: new insights from a decade of research. J Biosci. 2020; 45(1):1–14.

[22] Tahiliani M, Koh KP, Shen Y, Pastor WA, Bandukwala H, Brudno Y, Agarwal S, Iyer LM, Liu DR and Aravind L. Conversion of 5-methylcytosine to 5-hydroxymethylcytosine in mammalian DNA by MLL partner TET1. Science. 2009; 324(5929):930–935.

[23] Gao Y, Chen J, Li K, Wu T, Huang B, Liu W, Kou X, Zhang Y, Huang H and Jiang Y. Replacement of Oct4 by Tet1 during iPSC induction reveals an important role of DNA methylation and hydroxymethylation in reprogramming. Cell Stem Cell. 2013; 12(4):453–469.

[24] Olariu V, Lövkvist C and Sneppen K. Nanog, Oct4 and Tet1 interplay in establishing pluripotency. Sci Reports. 2016; 6(1):1–11.

[25] Di Stefano B, Sardina JL, van Oevelen C, Collombet S, Kallin EM, Vicent GP, Lu J, Thieffry D, Beato M and Graf T. C/EBPα poises B cells for rapid reprogramming into induced pluripotent stem cells. Nature. 2014; 506(7487):235–239.

[26] Hu X, Zhang L, Mao S-Q, Li Z, Chen J, Zhang R-R, Wu H-P, Gao J, Guo F and Liu W. Tet and TDG mediate DNA demethylation essential for mesenchymal-to-epithelial transition in somatic cell reprogramming. Cell Stem Cell. 2014; 14(4):512–522.

[27] Brumbaugh J, Di Stefano B and Hochedlinger K. Reprogramming: identifying the mechanisms that safeguard cell identity. Development. 2019; 146(23):1–17.

[28] Bryce WC, Markoulaki S, Beard C, Hanna J and Jaenisch R. Single-gene transgenic mouse strain for reprogramming adult somatic cells. Nat Methods. 2010; 7(1):56–9.

# 8

# Isolation and Expansion of Placental Stem Cells (P-MSCs)

**B. Shadpoor[1,2], L. S. Sherman[1,2], M. P. Romagano[3], Shauna F. Williams[3], K. Krishnamoorthy[3], K. Powell[3], Y. K. Kenfack[1,2], and P. Rameshwar[1]**

[1]Division of Hematology/Oncology, Department of Medicine, New Jersey Medical School, Rutgers Biomedical and Health Sciences, USA
[2]School of Graduate Studies, Biomedical Sciences Programs – Newark, Rutgers University, USA
[3]Department of Obstetrics, Gynecology and Women's Health, Rutgers New Jersey Medical School, Rutgers Biomedical and Health Sciences, USA
**Corresponding Author:** Shauna F. Williams, Department of Obstetrics, Gynecology and Women's Health, Rutgers New Jersey Medical School, USA
E-mail: williash@njms.rutgers.edu
**Disclaimer:** The authors have nothing to declare.

## Abstract

Preeclampsia (PE) is a pregnancy-associated disorder that is typically diagnosed by gestational hypertension, proteinuria, or end organ failure. Resident placental stem cells, resembling other tissue-derived mesenchymal stem cells (MSCs), are at the interface of the mother and fetus. The placental stem cells (P-MSCs) can have key roles in proper placentation. These roles can be affected during PE pathogenesis. Thus, it is critical to be able to expand P-MSCs from different regions of the placenta since this will allow for proper dissection of pathologies linked to pregnancy, such as PE. Standardization of the method is critical for scientists across the globe to compare their research studies for proper development of translational studies, including clinical trials. Here, we describe the method to isolate P-MSCs from different placental regions and describe the cell's characterization by phenotype and functional differentiation.

## 8.1 Introduction

Mesenchymal stem cells (MSCs) are adult stem cells that are capable of tri-lineage differentiation. MSCs display a fibroblast-/spindle-like morphology and are plastic adherent under standard culture conditions. These cells are found in various tissues such as bone marrow, adipose tissue, umbilical cord, dental pulp, and placenta [1]. Evidence of regenerative potential suggests that MSCs could be important for homeostatic repair mechanisms during daily tissue insult.

MSCs are considered immune-privileged cells because of their ability to cross allogeneic barriers without immune rejection [2]. Additionally, MSCs can be licensed as anti-inflammatory cells, which then home to areas of inflammation and serve as the conduits for direct repair [2]. As a result, MSCs are attractive therapeutic tools for treating various diseases, such as myocardial infarction, cartilage regeneration, asthma, and preeclampsia (PE).

MSCs from PE-derived placenta (PE P-MSCs) display increased production of proinflammatory cytokines, antiangiogenic factors, and inhibitory miRNAs that induce cell cycle arrest and apoptosis [3, 4]. Since these biological markers have been similarly observed in PE since the 1970s, PE P-MSCs have been implicated as major contributors to PE pathogenesis. Given that women's health has historically been overlooked in research and medicine, it is imperative that scientists expand their understanding of P-MSCs to improve perinatal outcomes and decrease pregnancy-related morbidity and mortality.

P-MSCs can be isolated from various layers of the placenta, including the chorionic membrane (CM), amniotic membrane (AM), chorionic villi (CV), decidua (D), and the umbilical cord (UC).

## 8.2 Reagents and Materials

### 8.2.1 Placenta sample

- See placental collection and preparation in Section 8.3.

### 8.2.2 Reagents

- PBS (1×) – (SIGMA Life Science d5652-1L: Dulbecco's phosphate buffered saline)
- Type II collagenase:

- MSC media:

  i. Dulbecco's Modified Eagle Medium (DMEM) – high glucose (ThermoFisher Scientific)

  ii. Fetal bovine serum (FBS) (Gibco – ThermoFisher Scientific)

  iii. 10 mg/ml penicillin–streptomycin (P-S) (Gemini Bioproducts)

  iv. 200 mM L-Glutamine (ThermoFisher Scientific)

  v. P-MSC media:

  - DMEM          90 ml
  - FBS            10 ml
  - P-S             1 ml
  - L-Glutamine    1 ml

## 8.2.3 Materials and surgical accessories

- Forceps
- Scissors
- 100 μm Sterile membrane

## 8.2.4 Equipment

- Centrifuge
- Flow cytometer

## 8.3 Method

### 8.3.1 Placental collection and preparation

- Depending on the study's focus, inclusion or exclusion criteria should be established when selecting placenta samples obtained from donors.

  i. Regulatory approval: The study must be approved by your Institutional Review Board. Although placenta is typically a discard tissue, the IRB will determine the requirements for a donor to sign an informed consent.

### 8.3.2 Handle the placentas under standard conditions in a laminar flow hood

1.  Placentas are collected after term vaginal or cesarean delivery and soaked in PBS

2.  Wash the placentas with sterile tissue culture grade PBS (pH 7.4) until most of the blood is removed by macroscopic examination (Note 1)

3.  Mechanically separate the CM, AM, CV, D, and UC layers (Note 2)

### 8.3.3 Mechanical separation

1.  Use forceps and scissors to mince the five tissue layers into 3–5 mm pieces

2.  Proceed to the explant method (see below) or perform the type-II collagenase digestion method (see below) (Note 3)

### 8.3.4 Explant method

1.  Place approximately 4–8 pieces of minced tissue into 100 mm² tissue culture plates containing 8 ml of MSC media (Note 4)

2.  Move the plates to a 37 °C, 5% $CO_2$ incubator for up to two weeks

3.  Visualize the plates with a microscope for extravasated P-MSCs (Note 5)

### 8.3.5 Type-II collagenase digestion method (Note 3)

1.  Treat minced tissue with 0.075% type-II collagenase; then leave the sample to incubate at 37 °C, 5% $CO_2$ for 1 hour

2.  Filter the digested tissue through a 100 μm sterile membrane (Note 6)

3.  Centrifuge the filtrate at 300 $g$ for 7 minutes at 25 °C

4.  Resuspend the pellet in P-MSC media

5.  Seed cells in a 6-well or 100 mm plate (plasma treated). The type of culture dish will depend on the total pellet size

6.  After three days, the adherent cells are morphologically spindle-shaped

    a. Replace 50% of the media with fresh media weekly until the cells reached 70%–80% confluence. The adherent cells from one dish is split into two separate culture dishes – use the original dish and one new dish (Note 7).

7.  Use P-MSCs after passage 3. If your quality control (see below for characterization) indicates that the P-MSCs are still multipotent, you may use up to passage 11 in the assays

### 8.3.6 MSC characterization

• Reference Method 15

### 8.3.6.1 Flow cytometry

At the third passage, phenotypic characterization of P-MSCs is conducted for the following surface markers:

• CD34⁻, CD45⁻, low-to-negative HLA-DR, CD73⁺, CD90⁺, and CD105⁺.

### 8.3.7 P-MSC multilineage differentiation

• Reference Method 15

1.  P-MSCs are tested for tri- or bi-lineage differentiation as follows (Note 8)

    a. Osteogenic differentiation potential is assessed by imaging cells with a von Kossa stain to detect osteoblasts.

    b. Adipogenic differentiation potential is assessed by imaging cells with an Oil Red O stain to detect adipocytes.

    c. Chondrogenic differentiation potential is assessed by imaging cells with a toluidine blue stain to detect chondrocytes.

## 8.4 Notes

Note 1: As a rule of thumb, we have found that three washes with large amounts of PBS are sufficient to remove most of the blood from the placental samples.

Note 2: If you are not experienced handling placental tissue, you are advised to consult an expert on placental anatomy. Mechanically separate the chorionic membrane (CM), amniotic membrane (AM), chorionic villi (CV), decidua (D), and the umbilical cord (UC).

Note 3: Both methods can efficiently isolate P-MSCs. However, the type-II collagenase digestion method, although rapid, results in lower P-MSC yield, as compared to the explant method.

Note 4: In cases where you are using the explant method, you can have better efficiency with MSC extravasation with smaller pieces of tissues. Thus, it is important to mince the tissue layers before placing them into 100 mm$^2$ tissue culture plates.

Note 5: At 48–72 hours after plating the tissues, the erythrocytes (RBCs) will lyse to generate debris. The can make visualization of the explanted cells difficult. Thus, before the MSCs have extravasated, it is recommended that you aspirate the media to clear debris and replace with fresh media.

Note 6: Regarding the type-II collagenase digestion method, you may place the minced tissue over a 100 μm sterile membrane that is placed on top of a 50 ml tube. Type-II collagenase is added to digest the tissue. After 30–60 minutes, the filtrate can then be seeded as per 3.5.1.5.

Note 7: To optimize MSC growth conditions while passaging cells, mix 50% of the old media with 50% fresh media. The old media contains various factors, including the MSC secretome, which helps maintain MSC integrity and function.

Note 8: When assessing the tri-/bi-lineage potential of P-MSCs, you may assess the cells for osteogenic, adipogenic, and chondrogenic differentiation potential. It is generally sufficient to test for two lineages. If your laboratory works with neurons, you may also use neuronal differentiation.

## References

[1] Barry FP, Murphy JM. Mesenchymal stem cells: clinical applications and biological characterization. Intl J Biochem Cell Biol. 2004;36:568–584.
[2] Le Blanc K, Tammik C, Rosendahl K, Zetterberg E, Ringden O. HLA expression and immunologic properties of differentiated and undifferentiated mesenchymal stem cells. Exp Hematol. 2003;31:890–896.
[3] Rolfo A, Giuffrida D, Nuzzo AM, Pierobon D, Cardaropoli S, Piccoli e, Giovarelli M, Todros T. Pro-inflammatory profile of preeclamptic placental mesenchymal stromal cells: new insights into the etiopathogenesis of preeclampsia. PloS One. 2013;8:e59403.
[4] Hwang JH, Lee MJ, Seok OS, Paek YC, Cho GF, Seol HJ, Lee JK, Oh MJ. Cytokine expression in placenta-derived mesenchymal stem cells in patients with pre-eclampsia and normal pregnancies. Cytokine. 2010;49:95–101.

# 9

# Expanding Human Mesenchymal Stem Cells from the Umbilical Cord

**Lauren S. Sherman[1,2], Andrew Petryna[1,2], Matthew P. Romagano[3], and Pranela Rameshwar[1]**

[1]Department of Medicine, Rutgers New Jersey Medical School, USA
[2]Rutgers School of Graduate Studies at New Jersey Medical School, USA
[3]Department of Obstetrics and Gynecology, Rutgers New Jersey Medical School, USA
**Corresponding Author:** Lauren Sherman, Rutgers New Jersey Medical School, USA
E-mail: shermala@njms.rutgers.edu
**Disclaimer:** The authors have no conflict to declare.

## Abstract

Umbilical cord (UC) is considered a discarded tissue. However, this cord is a "rich" source of mesenchymal stem cells (UC-MSCs). As compared to other sources from adipose and bone marrow, the UC has reduced ethical concern. Despite the ease of obtaining UC, you are most likely to require an approved informed consent by the mother. This chapter provides a stepwise procedure, adapted from a method provided by Dr. Francesco Dazzi. The technique describes the method to expand UC-MSCs from UC. Thus, UC-MSCs might not be able to replace similar cells from other sources. Thus, investigators will be required to determine the application in experimental studies.

## 9.1 Introduction

Umbilical cord (UC) has garnered much interest in the last two decades as an ethical source of primitive stem cells. The umbilical cord is otherwise a discard tissue, although consent from the mother is still required. Thus, UC-derived cells do not have the same ethical issues as cells sourced from

69

embryos or fetal tissue or the logistical issues stymying from primary tissues like bone marrow.

The vast majority of birth tissues such as placenta and UC are discarded as medical waste, making UC mesenchymal stem cells (UC-MSCs) a practical and readily available source of stem cells [1]. UC-MSCs are found in the Wharton's Jelly, a highly flexible but firm tissue surrounding the umbilical vessels that make up the bulk of the UC [2]. These stem cells, often called Wharton's Jelly/umbilical cord stem cells or stromal cells, have gained considerable attention since their initial isolation in the 1990s. Like many other sources of MSCs, UC-MSCs have been shown to have great proliferative potential with an ability to differentiate into osteogenic, adipogenic, chondrogenic, neuronal, and other various cell types [3–5]. Consistent with accepted international standards of MSC definition, they are also CD105, CD73, and CD29 positive [6]. There is considerable interest to use MSCs in cellular therapy. Thus, UC-MSCs have become an excellent source of stem cells for use in research, as off-the-shelf cells, and possible replacements for other cell MSC sources such as bone marrow. Here we will describe a method to isolate and culture these cells.

## 9.2 Method and Reagents

### 9.2.1 Procuring the umbilical cord (UC)

UC and placenta require processing immediately after collection. It is thus imperative that primary tissue samples collected and harvested as quickly as possible to ensure highest cell counts and viability. Although UC and placenta are considered as discarded products, it is imperative to remember that they may be required for pathology or other clinical care, which may delay – or even hamper – isolation of viable cells (e.g., tissue fixation). In such a case, a study-team member should accompany the sample to pathology to request a piece of tissue prior to fixation if this is amenable to the pathologist. While preferable to process the UC sample within several hours, it may be stored at 4 °C and processed for up to 24 hours.

Sample collection should be closely coordinated with the responsible personnel in the Ob/Gyn service. This point person will notify the study-team when delivery is imminent so that reagents can be prepared and ready for processing the UC. Depending on the institutional review board (IRB), obtain the consent from the mother. Place the UC and any other tissues needed for research into a sterile bag or container for transport to the laboratory. If UC blood is required, this should be collected prior to storing and transporting the UC – ensure heparin is placed in the collection tube to prevent coagulation of the blood.

If the tissue will not be processed immediately, the tissue should be stored in a saline solution (e.g., tissue culture grade PBS, pH 7.4) supplemented with antibiotics. Care in handling the samples will allow for isolation of UC-MSCs with little contamination from human umbilical vascular endothelial cells (HUVECs) that are within the UC vein. Depending on your experimental question, it is recommended to take UC from otherwise healthy subjects.

### 9.2.2 Reagents

All reagents should be sterile.

1.  DMEM (high glucose) supplemented with 2 mM L-glutamine, and sodium bicarbonate

2.  Gentamicin, 500 mM stock solution recommended

3.  Fetal bovine serum, heat inactivated at 56 °C for 45 minutes

4.  Tissue culture grade phosphate buffered saline (PBS), calcium and magnesium free

5.  100 mm round, plasma-gas-treated sterile cell tissue culture plates

6.  150 mm round, sterile dishes (may be non-tissue culture treated)

7.  10 cc Syringes

8.  16-Gauge hypodermic needles

9.  Scalpel, blade, and/or scissors

10. Forceps

11. Optional: sterile gloves

12. 0.05% Trypsin–EDTA solution

### 9.2.3 Tissue culture

All steps are done under a sterile laminar flow hood applicable for cell culture. During the initial steps, the UC can be moved and manipulated with sterile forceps. If the UC is to be manipulated with the hands rather than forceps, sterile or surgical gloves should be used.

1.  Prepare complete culture media with antibiotic:

    a.  Complete isolation media:

        i.  DMEM (high glucose) supplemented with 2 mM L-glutamine, and sodium bicarbonate.

        ii.  Gentamicin, final concentration 50 µg/ml.

        iii.  Fetal bovine serum, heat inactivated, and final concentration 10%.

    b.  Complete expansion media:

        i.  DMEM (high glucose) supplemented with 2mM L-glutamine, and sodium bicarbonate.

        ii.  Fetal bovine serum, heat inactivated, and final concentration 10%.

        iii.  Optional: whatever antibiotic(s) are traditionally used in the lab's tissue culture protocols.

    c.  Antibiotic-supplemented PBS:

        i.  PBS.

        ii.  Gentamicin, final concentration 50 µg/ml.

2.  Transfer the UC-containing transport vessel to a laminar flow cell culture hood. If the sample was not transferred in antibiotic-supplemented PBS, transfer to a sterile container with this if not already. When ready to work with the tissue, transfer the UC onto a sterile surface (e.g., a 150 mm dish) where the UC can be easily manipulated. Transfer can be done with forceps or a sterile-gloved hand.

3.  Load a 10 cc syringe with antibiotic-supplemented PBS and use the 16-guage needle to flush residual blood from the umbilical veins and arteries.

4.  Cut the UC into 2–5 cm long pieces of Wharton's Jelly. If possible, excise the arteries and veins; alternatively cut thin strips of the Wharton's Jelly that do not contain the major vasculature. Slide the pieces longitudinally to expose the inner surface. Mince these pieces into smaller fragments (approximately 2 mm) (Note 1).

5.  Place the fragments into a tissue-culture-treated 100 mm sterile dish. Allow them to dry on the surface for approximately 15 minutes prior to

adding media. This allows the tissue to adhere to the plate, enhancing cell extravasation from the tissue.

6.   Carefully add 10 ml of complete isolation media to the plate.

7.   Incubate undisturbed at 37 °C in a humidified atmosphere with 5% $CO_2$ for 6–7 days. As days 6–7 approached, the plate may be checked for cell establishment surrounding pieces of tissue. Do not disturb the plates any more than necessary lest you disturb the tissue.

8.   After 6–7 days, remove the tissue fragments. UC tissue can be removed by gently rinsing the plate twice with PBS.

9.   Add 10 ml of complete expansion medium and allow the cells to proliferate for an additional 7–14 days, conducting a partial media change – replacing approximately 50% of the culture media with complete expansion media – every 3–4 days or if the media should turn yellow.

10.   Once the cells reach 70% confluency, they should be sub-cultured. If multiple sites of high density are noted, the cells can be trypsinized and redistributed on the same plate to encourage further proliferation.

11.   See Chapters 3 and 16 for culture and cryopreservation of MSCs.

12.   At passage 3, conduct quality control on your cells. This should include phenotyping for MSC markers. The panel of antibodies should include CD31 and CD45 to eliminate endothelial cells and hematopoietic cells, respectively. Refer to Chapter 16 for the method on phenotyping of MSCs (see Chapter 16). As positive control for CD31, you could use umbilical endothelial cells.

## 9.3 Notes

Note 1: The UC is slippery. If it is too slippery to handle, wait several minutes for the outside to dry – this might help with grip.

## References

[1]   Ding D-C, Chang Y-H, Shyu W-C and Lin S-Z. Human umbilical cord mesenchymal stem cells: a new era for stem cell therapy. Cell Transpl. 2015; 24(3):339–347.

[2]   Ghosh K, Ghosh S and Gupta A. Tensile properties of human umbilical cord. The Indian J Med Res. 1984; 79:538–541.

[3] Karahuseyinoglu S, Cinar O, Kilic E, Kara F, Akay GG, Demiralp DÖ, Tukun A, Uckan D and Can A. Biology of stem cells in human umbilical cord stroma: in situ and in vitro surveys. Stem Cells. 2007; 25(2):319–331.

[4] Kobayashi K, Kubota T and Aso T. Study on myofibroblast differentiation in the stromal cells of Wharton's jelly: expression and localization of α-smooth muscle actin. Early Human Dev. 1998; 51(3):223–233.

[5] Mitchell KE, Weiss ML, Mitchell BM, Martin P, Davis D, Morales L, Helwig B, Beerenstrauch M, Abou-Easa K and Hildreth T. Matrix cells from Wharton's jelly form neurons and glia. Stem cells. 2003; 21(1):50–60.

[6] Wang HS, Hung SC, Peng ST, Huang CC, Wei HM, Guo YJ, Fu YS, Lai MC and Chen CC. Mesenchymal stem cells in the Wharton's jelly of the human umbilical cord. Stem Cells. 2004; 22(7):1330–1337.

# 10

# Laboratory Isolation of Leukemia Stem Cells

## Michael Haddadin and Shyam A. Patel

Department of Medicine, Division of Hematology/Oncology, University of Massachusetts Chan Medical School, USA
**Corresponding Author:** Shyam A. Patel, University of Massachusetts Chan Medical School, USA
E-mail: shyam.patel@umassmed.edu
**Disclaimer:** The authors have nothing to declare.

## Abstract

Leukemia is a group of heterogeneous blood disorders that involve clonal proliferation and developmental arrest of mature or immature blood cells. The leukemic cells can interfere with the production of normal blood cells and cause a variety of complications. Different diagnostic methods have evolved over the past century for the diagnosis of leukemia and the detection of measurable residual disease (MRD) and to assess the response to treatment. MRD is the subset of cells that persist in the bone marrow after treatment. Together, MRD and leukemia stem cell (LSC) persistence are believed to serve as the strongest risk factors for leukemia relapse. Flow cytometry, next-generation sequencing, and polymerase chain reaction are the most commonly used methods to identify these LSCs. Flow cytometry offers a unique ability to simultaneously assess and correlate multiple phenotypic properties at the single-cell level in a timely and efficient manner. Application of this technique to the detection of residual acute leukemia after therapy has proven its importance to monitor response to therapy and provide prognostic information. In this protocol, the flow cytometric detection of MRD and LSCs is outlined herein, with an emphasis on the leukemia-associated immunophenotype (LAIP) and LSC markers.

## 10.1 Introduction

Leukemia is triggered by hematopoietic progenitor cells in the bone marrow that become mutated and clonally expanded into leukemic blasts, which do not fully differentiate normally functioning blood cells. Leukemia can be sub-categorized based on several factors as follows: rate of disease progression, namely acute (rapid, within weeks to months) or chronic (slow, within months to years) and the type of malignant cells, either originating from the lymphoid or myeloid lineage [1, 2]. Acute myeloid leukemia (AML) is the most common adult leukemia. In the United States, the incidence is reported as 3–5 cases per 100,000 population with a five-year survival rate ranging between 10% and 30% depending on multiple factors [3, 4].

Numerous advances had occurred in diagnosing and treating AML. The ability to isolate and identify leukemia cell lines serves as a rich resource of abundance to better understand the pathophysiology of hematopoietic tumors. Leukemia cells are now universally used and have become irreplaceable tools in a multitude of research areas.

Advances in AML treatment, for example, have considerably improved initial response rates and patients' quality of life, though survival rates remain low, largely due to disease relapse [5]. A subset of cells persists and is resistant to induction treatment. This is known as measurable residual disease (MRD), and it is a well-established factor for AML relapse [6–8]. Then MRD subset has a smaller population of therapy-resistant leukemic stem cells (LSCs). This is suggested to be at the base of disease and serves as the source of outgrowth of residual cells to relapse, and the frequency of these LSCs is also of prognostic significance [9–11]. The term LSC was first introduced by John Dick's laboratory in 1994, when it was demonstrated that only the leukemic cells expressing the same markers as normal adult hematopoietic stem cells (CD34$^+$CD38$^-$) could initiate hematopoietic malignancy [12, 13]. There is increasing evidence that LSCs are clinically significant, as they possess capabilities of self-renewal, proliferation, and differentiation [11, 13, 14].

Methods for evaluating MRD in AML are evolving and await standardization [7]. The most commonly used methods for monitoring MRD include reverse-transcriptase polymerase chain reaction (RT-PCR) and multiparameter flow cytometry (MFC). Some institutes use next-generation sequencing (NGS) to detect gene mutations for MRD analysis [15]. Each method has associated benefits and limitations [6, 15–17]. An ideal MRD assay would be sensitive to low levels of MRD and suitable for frequent analysis. Future residual disease assessment might benefit from the combined use of NGS and MFC. NGS has the highest sensitivity and can detect one in $10^5$–$10^6$ aberrant cells [18, 19]. Flow cytometry has a sensitivity and can detect one in $10^3$–$10^4$

aberrant cells in MRD [20, 21]. Further improvement of predicting an AML relapse is accomplished by the incorporation of the LSC frequency [22]. The prognostic implication of MRD, comparison of MRD evaluation methods and assay, LSC phenotypes, and the relation between MRD and LSC are beyond the scope of this chapter.

The ability to detect MRD by flow cytometry for patients with acute leukemia is well-described in the literature. The basic method relies on the discovery of antigens that are differentially expressed between leukemic populations and their normal counterparts in bone marrow, i.e., informative antigens [20, 21]. This aberrant expression is known as the leukemia-associated immunophenotype (LAIP). The use of combinations of monoclonal antibodies against these antigens, in concert with reagents that allow recognition of specific hematopoietic cell lineages and maturational stages, allows for the detection of immunophenotypic abnormalities on discrete subpopulations and is key to the success of the method. Neoplastic or leukemic populations are recognized by an immunophenotype that is different or deviates from that seen on normal populations. It is the immunophenotype of the normal populations that remain relatively constant and serve as reference points for population identification and assessment [20, 22–24]. We hereby describe detection and isolation of $CD34^+CD38^-$ LSCs using MFC.

## 10.2 Materials

### 10.2.1 Flow cytometers

Flow cytometers have five main components: a flow cell, a measuring system, a detector, an amplification system, and a computer for analysis of the signals. Acquisition is the process of collecting data from examined samples. In leukemia and various other hematological malignancies, the acquisition of cells is based on fluorescent-labeled antibodies.

### 10.2.2 Reagents

 i. PBS/BSA/azide (phosphate buffered saline with 0.3% BSA and 0.1% sodium azide).

 ii. Buffered $NH_4Cl$ lysing solution containing 0.25% formaldehyde.

### 10.2.3 Monoclonal antibody panels

 i. Panels of monoclonal antibodies (mAbs) that are suitable for identifying LSCs can vary depending on the manufacturer, evaluation at the time of diagnosis versus MRD and cell of origin. IgG1 reagents are preferred to minimize the potential for complement-mediated antibody interactions [25]. Most panels have two tubes with different mAbs attached to fluorochromes. Examples of fluorochromes are fluorescein isothiocyanate (FITC), phycoerythrin (PE), PerCP-Cy5.5 (PCP5.5), PE-Cy7 (PC7), allophycocyanin (APC), Pacific blue (PB), and APC-Cy7 (APC7).

ii. One example of LSC antibody panel includes: FITC: CD45RA, PerCP-CY5.5: CD123, PC7: CD34, APC: CD38, APC7: CD44, BV421: CD33, KO: CD45, and many others. The panel for the detection of residual acute myeloid leukemia includes: HLA-DR, CD15, CD33, CD19, CD117, CD13, CD38, CD34, CD71, and CD45.

## 10.2.4 Sample

A 10 ml bone marrow in heparin (e.g., lithium heparin, 102 IU coated tubes; Becton Dickinson) or EDTA tube (e.g., plastic K2EDTA 7.2 mg; Becton Dickinson).

## 10.3 Methods

### 10.3.1 Preparation of the bone marrow (BM) sample

i. Assess concentration of white blood cells (WBCs) in the BM sample and define the needed volume of WBC.

ii. Add lysing solution to the tube containing white blood cell suspension to lyse any red blood cells. Mix gently by inverting the tubes and incubate for 10 minutes at room temperature and then centrifuge for 7 minutes at 700 $g$.

iii. Remove supernatant and resuspend cell pellet in excess PBS. Centrifuge for 7 minutes at 700 $g$.

iv. Remove supernatant and resuspend cell pellet in PBS to a cell concentration of ~100 × 10$^6$ WBC/ml before dividing cell suspension evenly over the four different FACS tubes.

### 10.3.2 Staining of WBC

i. Add the mAbs into different tubes. Mix gently and incubate cell suspensions (20 µl) with the appropriate antibodies (20 µl premix containing all eight antibodies) for 15 minutes at room temperature while protecting from light.

ii. Add 3 ml PBS per tube to wash the stained cells. Centrifuge cells at 400× g for 5 minutes (with brake). Remove supernatant and resuspend cell pellet in 300 μl PBS.

### 10.3.3 Flow cytometry LSC assessment

Use a flow cytometer to measure as many gated WBCs as possible, at least $4 \times 10^6$, especially for follow-up samples. High WBC counts are recommended to enable proper LSC detection.

### 10.3.4 Gating and data analysis

One basic strategy used as a starting point in histograms to identify a specific cell population is the use of lineage-associated antigens in a combination with side scatter. B cells of all types may be well-identified using CD19 in combination with side scatter; other markers can help further identify the maturation level of the B-cell subpopulation of interest. For immature populations in general, the use of CD45 versus side scatter is usually the starting point, to allow the identification of progenitors characteristically having lower CD45.

i. Using the appropriate software and gating system to assess leukemia-associated immunophenotypes (LAIPs)

ii. Attempt identifying the WBC using CD45 positive cells. Within the population of WBC, gate the cells with intermediate CD45 and have low side scatter for immature cells.

iii. Gate CD34$^+$CD38$^-$ cells and identify them in a plot.

iv. Use the mAb markers above to identify presumed normal hematopoietic stem cell from LSC

## 10.4 Notes

Note 1: Note for simultaneous labeling of white cells with mAbs against the surface and cytoplasmic antigen; antibodies against cytoplasmic antigens require careful titration to optimize the signal-to-noise ratio and often must be used at sub-saturating concentrations to minimize background. It needs antibody panels lacking reagents directed against cytoplasmic antigens [26].

Note 2: Prolonged fixation and exposure to fluorescent light may degrade Cy7-containing tandem fluorochromes.

Note 3: For diagnostic leukemia-associated immunophenotypes and LSC determination, peripheral blood can be used because high percentages of blasts are often present in the peripheral blood at the time of diagnosis. However, for follow-up MRD and LSC assessment, BM is preferred because the frequency of leukemic cells is usually lower in peripheral blood compared to BM [20, 23, 26].

Note 4: For an assay where a detection sensitivity of 0.01% of white cells is desired, this will require the acquisition of 1,000,000 total white cell events. The aim for MRD assessment is to acquire 100,000 events for diagnosis samples and 1,000,000 events for follow-up samples [23, 26].

## References

[1] Arber DA, Orazi A, Hasserjian R, Thiele J, Borowitz MJ, Le Beau MM, Bloomfield CD, Cazzola M and Vardiman JW. The 2016 revision to the World Health Organization classification of myeloid neoplasms and acute leukemia. Blood. 2016; 127(20):2391–2405.

[2] Swerdlow SH, Campo E, Pileri SA, Harris NL, Stein H, Siebert R, Advani R, Ghielmini M, Salles GA, Zelenetz AD and Jaffe ES. The 2016 revision of the World Health Organization classification of lymphoid neoplasms. Blood. 2016; 127(20):2375–2390.

[3] Dores GM, Devesa SS, Curtis RE, Linet MS and Morton LM. Acute leukemia incidence and patient survival among children and adults in the United States, 2001-2007. Blood. 2012; 119(1):34–43.

[4] Siegel RL, Miller KD and Jemal A. Cancer Statistics, 2017. CA: a cancer journal for clinicians. 2017; 67(1):7–30.

[5] Oliva EN, Franek J, Patel D, Zaidi O, Nehme SA and Almeida AM. The Real-World Incidence of Relapse in Acute Myeloid Leukemia (AML): A Systematic Literature Review (SLR). Blood. 2018; 132(Supplement 1): 5188–5188.

[6] Jaso JM, Wang SA, Jorgensen JL and Lin P. Multi-color flow cytometric immunophenotyping for detection of minimal residual disease in AML: past, present and future. Bone Marrow Transpl. 2014; 49(9):1129–1138.

[7] Schuurhuis GJ, Heuser M, Freeman S, Béné MC, Buccisano F, Cloos J, Grimwade D, Haferlach T, Hills RK, Hourigan CS, Jorgensen JL, Kern W,

Lacombe F, et al. Minimal/measurable residual disease in AML: a consensus document from the European LeukemiaNet MRD Working Party. Blood. 2018; 131(12):1275–1291.

[8] Hokland P and Ommen HB. Towards individualized follow-up in adult acute myeloid leukemia in remission. Blood. 2011; 117(9):2577–2584.

[9] Becker MW and Jordan CT. Leukemia stem cells in 2010: current understanding and future directions. Blood Rev. 2011; 25(2):75–81.

[10] Bonnet D and Dick JE. Human acute myeloid leukemia is organized as a hierarchy that originates from a primitive hematopoietic cell. Nature Med. 1997; 3(7):730–737.

[11] Eppert K, Takenaka K, Lechman ER, Waldron L, Nilsson B, van Galen P, Metzeler KH, Poeppl A, Ling V, Beyene J, Canty AJ, Danska JS, Bohlander SK, et al. Stem cell gene expression programs influence clinical outcome in human leukemia. Nature Med. 2011; 17(9):1086–1093.

[12] Lapidot T, Sirard C, Vormoor J, Murdoch B, Hoang T, Caceres-Cortes J, Minden M, Paterson B, Caligiuri MA and Dick JE. A cell initiating human acute myeloid leukaemia after transplantation into SCID mice. Nature. 1994; 367(6464):645–648.

[13] Zeijlemaker W, Grob T, Meijer R, Hanekamp D, Kelder A, Carbaat-Ham JC, Oussoren-Brockhoff YJM, Snel AN, Veldhuizen D, Scholten WJ, Maertens J, Breems DA, Pabst T, et al. CD34(+)CD38(-) leukemic stem cell frequency to predict outcome in acute myeloid leukemia. Leukemia. 2019; 33(5):1102–1112.

[14] Ishikawa F, Yoshida S, Saito Y, Hijikata A, Kitamura H, Tanaka S, Nakamura R, Tanaka T, Tomiyama H, Saito N, Fukata M, Miyamoto T, Lyons B, et al. Chemotherapy-resistant human AML stem cells home to and engraft within the bone-marrow endosteal region. Nature Biotechnol. 2007; 25(11):1315–1321.

[15] Kohlmann A, Grossmann V, Nadarajah N and Haferlach T. Next-generation sequencing - feasibility and practicality in haematology. Br J Haematol. 2013; 160(6):736–753.

[16] Paietta E. Consensus on MRD in AML? Blood. 2018; 131(12):1265–1266.

[17] Kern W, Haferlach C, Haferlach T and Schnittger S. Monitoring of minimal residual disease in acute myeloid leukemia. Cancer. 2008; 112(1):4–16.

[18] Thol F, Gabdoulline R, Liebich A, Klement P, Schiller J, Kandziora C, Hambach L, Stadler M, Koenecke C, Flintrop M, Pankratz M, Wichmann M, Neziri B, et al. Measurable residual disease monitoring by NGS before allogeneic hematopoietic cell transplantation in AML. Blood. 2018; 132(16):1703–1713.

[19] Jongen-Lavrencic M, Grob T, Hanekamp D, Kavelaars FG, Al Hinai A, Zeilemaker A, Erpelinck-Verschueren CAJ, Gradowska PL, Meijer R, Cloos J, Biemond BJ, Graux C, van Marwijk Kooy M, et al. Molecular Minimal Residual Disease in Acute Myeloid Leukemia. New Engl J Med. 2018; 378(13):1189–1199.

[20] Grimwade D and Freeman SD. Defining minimal residual disease in acute myeloid leukemia: which platforms are ready for "prime time"? Blood. 2014; 124(23):3345–3355.

[21] Al-Mawali A, Gillis D, Hissaria P and Lewis I. Incidence, sensitivity, and specificity of leukemia-associated phenotypes in acute myeloid leukemia using specific five-color multiparameter flow cytometry. Am J Clin Pathol. 2008; 129(6):934–945.

[22] Terwijn M, Zeijlemaker W, Kelder A, Rutten AP, Snel AN, Scholten WJ, Pabst T, Verhoef G, Löwenberg B, Zweegman S, Ossenkoppele GJ and Schuurhuis GJ. Leukemic stem cell frequency: a strong biomarker for clinical outcome in acute myeloid leukemia. PloS One. 2014; 9(9):e107587.

[23] Zeijlemaker W, Kelder A, Cloos J and Schuurhuis GJ. Immunophenotypic Detection of Measurable Residual (Stem Cell) Disease Using LAIP Approach in Acute Myeloid Leukemia. Curr Protocols Cytometry. 2019; 91(1):e66.

[24] Schuurhuis GJ, Ossenkoppele GJ, Kelder A and Cloos J. Measurable residual disease in acute myeloid leukemia using flow cytometry: approaches for harmonization/ standardization. Expert Rev Hematol. 2018; 11(12):921–935.

[25] Wood BL and Levin GR. Interactions between mouse IgG2 antibodies are common and mediated by plasma C1q. Cytometry Part B, Clin Cytometry. 2006; 70(5):321–328.

[26] Wood BL. Flow cytometric monitoring of residual disease in acute leukemia. Methods in Mol Biol. 2013; 999:123–136.

# 11

# Isolation and Characterization of Cancer Stem Cells from Glioblastoma Multiforme and Breast Cancer Cell Lines

**Vibha Harindra Savanur[1,2], Alejandra I. Ferrer[1,2], and Pranela Rameshwar[1]**

[1]Department of Medicine, Rutgers New Jersey Medical School, USA
[2]Rutgers School of Graduate Studies at New Jersey Medical School, USA

**Grant:** This work was supported by the following grants: New Jersey Commission on Cancer Research (NJCCR) Predoctoral Fellowship to A.I.F. (DCSH20PPC041) and MetaVivor Foundation.
**Corresponding Author:** Pranela Rameshwar, Rutgers New Jersey Medical School, USA
E-mail: rameshwa@njms.rutgers.edu
**Disclaimer:** The authors have nothing to declare.

## Abstract

Cancer stem cells (CSCs) are a major contributor for cancer relapse. CSCs are tumor-initiating cells that exhibit properties similar to healthy stem cells. CSCs evade treatment and immune detection by employing mechanisms such as dormancy and drug resistance, including upregulation of multi-drug resistance genes. Therefore, it is imperative to develop strategies that successfully eliminate CSCs to prevent cancer relapse. One of the major drawbacks in current research is the efficient isolation of CSCs. The development of methods that improve the isolation and characterization of CSCs is crucial to effectively target the cells without harming endogenous stem cells. Here, we describe the isolation and characterization of CSCs by using two reporter vector systems, Oct4a-GFP and SORE6-GFP, in breast cancer and glioblastoma, respectively. In addition, we describe methods to confirm the characterization of long-term CSCs such as *in vitro* tumorsphere assay and *in vivo* serial dilution. Overall, the methods depicted serve as effective ways to identify and isolate CSCs potentially facilitating the targeting and elimination of CSCs.

## 11.1 Introduction

Cancer stem cells (CSCs) are tumor-initiating cells that display properties similar to healthy stem cells and are the major source of cancer relapse [1]. CSCs can evade treatment and immune responses by undergoing dormancy, a process characterized by cell-cycle quiescence, which allows the cells to persist for extended periods [2]. As a self-preservation mechanism, CSCs express high levels of multi-drug resistance genes (i.e., ABC transporters) contributing to their survival after rigorous treatments [3, 4]. Therefore, it is important to develop treatment strategies that will circumvent the challenges posed by CSCs to successfully eradicate cancer and avoid relapse. In this chapter, we will address different techniques used to isolate and characterize CSCs from two types of cancer, breast cancer (BC) and glioblastoma multiforme (GBM).

One of the major challenges in the area of cancer research is to have a standardized method to isolate and characterize CSCs. These stem cells were described for leukemia and then identified in solid tumors as $CD44^+$/$CD24^{(-/low)}$/Lineage$^{(-)}$ [5, 6]. Subsequently, expressions of markers such as CD133 and aldehyde dehydrogenase have also been used to identify CSCs from different tumors including GBM and BC [7–9]. In addition, CSCs display a high expression of core stem cell genes, such as Octamer-4a (Oct4a), Nanog, Sox2, and Klf-4 [2]. However, these markers are commonly expressed by healthy stem and progenitor cells, as well as some neural and epithelial cells [10, 11]. Hence, research should be focused on determining factors that will solely target CSCs without harming the healthy stem cells.

CSCs undergo asymmetric division, a property that allows their self-renewal or differentiation into cancer progenitors [3]. The ability of CSCs to form progenitors is measured *in vitro* by performing the tumorsphere assay [12, 13]. A cancer cell is classified as a CSC if it has the capacity to form spheres after dissociation of the tumor and single cell passaging. *In vivo*, the tumor-initiation properties of CSCs are measured by injecting a limited number of cells into the dorsal flank of immune deficient mice [13]. As part of this chapter, we will describe *in vitro* and *in vivo* methods to validate CSC stemness. Additionally, we will provide strategies to isolate CSCs, including the use of two vector reporter systems, Oct4a-GFP and Sox2-Oct4a response element (SORE)-GFP, which allows the identification and collection of CSCs based on their GFP expression by flow cytometry [13, 14].

Lastly, we will discuss the selection of CSCs induced by chemotherapy exposure [15].

## 11.2 Materials

*Isolation of CSCs by fluorescence activated cell sorting (FACS):*

### 11.2.1 Reagents for isolation of Oct4a-GFP reporter vector from transformed bacteria

   i.  Luria Broth Powder (ThermoFisher, Cat. No. 12795027)

  ii.  Kanamycin Powder (Carolina, Cat. No. 216881)

 iii.  QIAprep Spin Miniprep Kit (QIAGEN, Cat. No. 27106)

### 11.2.2 Reagents for isolation of SORE6-GFP reporter vector from transformed bacteria

   i.  Luria Broth Powder (ThermoFisher, Cat. No. 12795027)

  ii.  Ampicillin Powder (ThermoFisher, Cat. No. 11593027)

 iii.  QIAprep Spin Miniprep Kit (QIAGEN, Cat. No. 27106)

### 11.2.3 DNA quantification

   i.  NanoDrop spectrophotometer such as QIAxpert System Kit (QIAGEN, Cat. No. 9002340)

### 11.2.4 Reagents for transfection of BC cell lines

   i.  Lipofectamine 3000 Transfection Reagent (Invitrogen, Cat. No. L300015)

  ii.  Opti-MEM medium (Gibco, Cat. No. 11058021)

### 11.2.5 Reagents for lentiviral packaging of SORE6-GFP intended for transduction of GBM cell lines

   i.  HEK293T cells

  ii.  3rd Generation Packaging System Mix (ABM Inc.)

iii.  Lentiviral sample DNA

iv.  TransIT-Lenti Transfection Reagent (Mirus, Cat. No. MIR6603)

v.  0.45  m PVDF filter

vi.  Lenti-X Concentrator (Takara Biosciences Cat. No. 631232)

vii.  p24 Rapid Titer Kit (Takara Biosciences)

### 11.2.6  Reagents for transduction of GBM cell lines

i.  SORE6-GFP virus

### 11.2.7  Selection of Oct4a-GFP positive cells

i.  Geneticin (G418) antibiotic (Thermo Scientific, Cat. No. 10131035)

ii.  Cloning cylinder (Thermo Scientific, Cat. No. 09-552-20)

### 11.2.8  Selection of SORE6-GFP positive cells

i.  Puromycin antibiotic (Sigma-Aldrich, Cat. No. P9620)

ii.  Cloning cylinder (Thermo Scientific, Cat. No. 09-552-20)

### 11.2.9  Validation of transfection/transduction

i.  Fluorescence microscope to assess GFP intensity such as EVOS FL Auto 2 Imaging System (Thermo Scientific, Cat. No. AMAFD2000)

### 11.2.10  Reagents for sorting buffer used for sample preparation

i.  Hanks Balanced Salt Solution (HBSS) ($Ca2+/Mg2+$ Free)

ii.  1mM EDTA 25 mM

iii.  HEPES pH7.0

iv.  1% BSA

### 11.2.11  Sorting of CSC-enriched cells based on Oct4a-GFP or SORE6-GFP expression by flow cytometry

i.  1× phosphate saline buffer (PBS)

ii. 4% bovine serum albumin

*Chemotherapy selection of CSCs from heterogeneous populations:*

### 11.2.12 Carboplatin (Thermo Scientific, Cat. No. J60433.03)

### 11.2.13 *In vitro* serial dilution and tumorsphere assay

i. Neurobasal media (ThermoFisher, Cat. No. 21103049)

ii. B27 supplement (ThermoFisher, Cat. No. 17504044)

iii. EGF

iv. FGF

*In vivo serial dilution of CSCs:*

### 11.2.14 Female immune deficient BALB/cJ mice (The Jackson Laboratory, Strain No. 000651)

### 11.2.15 General materials for in vivo serial dilution

i. Matrigel Basement Membrane Matrix (Corning, Cat. No. 356237)

ii. Caliper

iii. 1 × PBS

iv. 1 ml syringe (BD, Cat. No. 309628)

v. 25-gauge needle (BD, Cat. No. 305125)

## 11.3 Methods

*Isolation of CSCs by fluorescence activated cell sorting (FACS):*

### 11.3.1 DNA isolation from transformed bacteria with Oct4a-GFP/SORE6-GFP vector

i. Sterilize benchwork by using 70% ethanol and 10% bleach. Carefully turn on the Bunsen burner.

ii. Add 3 ml of pre-sterilized Luria Broth (Miltenyi Biotec, Cat. No. 12795027) into a 15 ml round-bottom polypropylene tube.

iii. Add kanamycin/ampicillin to a final concentration of 50 µg/ml into the 15 ml round-bottom tube.

iv. Add 2 µl of transformed bacteria cryopreserved in glycerol into the 15 ml tubes.

v. Close the tubes and incubate the bacteria overnight in a shaker at 42 °C.

vi. After incubation, perform DNA isolation from bacteria with the QIAprep Spin Miniprep Kit as per the manufacturer's protocol.

vii. Quantitate DNA with a nanodrop spectrophotometer.

### 11.3.2 Transfection of breast cancer cells (BCCs)

i. Seed cells at 70% confluency on a 6-well plate and incubate overnight at 37 °C.

ii. Label two 1.5 ml centrifuge tubes as A and B, respectively, and add 125 µl of Opti-MEM medium.

iii. Add 5 µl of Lipofectamine 3000 reagent into tube A and vortex briefly.

iv. Into tube B, add 5 µl of P3000 reagent and 2.5 µg of DNA. Make sure to mix well the sample.

v. Add tube B components into tube A and mix carefully. Incubate DNA and lipid complex for 15 minutes at room temperature.

vi. Add components to the cells dropwise and incubate for 2–3 days to allow transfection to occur.

vii. Verify transfection efficiency by quantifying GFP intensity with a fluorescence microscope.

### 11.3.3 Selection of Oct4a-GFP positive cell clones

i. Treat cells with 500–700 µg/ml of G418 every 2–3 days until discrete cell colonies are formed.

ii. Once clones are observed and are sparse across the well, put a cylinder around the colony of interest, add 100 µl of trypsin and incubate for 5 minutes at 37 °C.

iii. Collect the cells, seed them into to a 96-well plate, and continue treating with 500–700 µg/ml of G418. Continue expanding cells (~15–20 × $10^6$ cells or more) for FACS application.

### 11.3.4 Transduction of GBM cell lines

i. Culture HEK293T cells in 100 mm dishes.

ii. Add 5 μg of Viral packaging DNA, 5 of sample DNA, 30 μl TransIT-lenti reagent, 1 ml of Opti-MEM media and incubate for 10 minutes at room temperature (RT).

iii. Add dropwise with shaking to the 100 mm dishes containing HEK cells.

iv. Incubate for 48 hours.

v. Collect supernatant and filter using a 0.45 μm PVDF filter.

vi. Concentrate Virus using Lenti-X Concentrator (Takara Biosciences) as per manufacture protocol.

vii. Viral particles were quantified using the p24 rapid titer kit.

viii. Aliquot the virus into low binding tubes and store at −80 .

### 11.3.5 Transduction of GBM cell lines with SORE6-GFP virus

i. Seed and culture 150,000 GBM cells in a 6-well plate.

ii. Add 50MOI virus dropwise when cells are at between 30% and 40% confluency.

iii. Change media after 48 hours and check under microscope for GFP+ cells.

### 11.3.6 Selection of clones to create GBM SORE6 cell lines

i. Add increasing doses of Puromycin to the transduced 6-well plate to select a single clone of GFP positive cell.

ii. Once the clone has formed a single colony, put a cylinder around the colony of interest, add 100 μl of trypsin, and incubate for 5 minutes at 37 °C.

iii. Collect the cells and expand. Freeze cells and store in liquid nitrogen.

### 11.3.7 Sorting of CSCs (GFP$^{high}$) from Oct4-aGFP (BC-CSCs) and SORE6-GFP (GBM-CSCs)

i. Wash and trypsinize cells and collect them in a tube.

ii.  Centrifuge cells at 300 *g* for 10 minutes.

iii.  Remove supernatant and resuspend cells sorting in 1–2 ml sorting buffer.

iv.  Perform FACS by running the sample along with appropriate controls (non-GFP cells, DAPI) and set the gates – GFP-high, GFP-medium, and GFP-low.

v.  Select each gate and collect the respective group of cells into collection tubes coated with BSA (long-term CSCs found in the GFP-high population).

vi.  Spin cells down at 500 *g* for 20 minutes.

vii.  Remove supernatant and resuspend in DMEM or other appropriate media and plate them in a 6-well plate and incubate.

### 11.3.8 In vitro serial dilution

### 11.3.8.1 Tumorsphere assay

i.  Set up the tumorsphere assay within 24–48 hours after sorting.

ii.  Wash, trypsinize, and collect the GFP$^{high}$ group of cells and count using trypan blue.

iii.  Dilute the cells using tumorsphere media and add 1 cell/well in a low attachment 96-well plate.

iv.  Add PBS on the outer most wells of the 96-well plate (protects sample from evaporation).

v.  Place in incubator and image under brightfield until a sphere is formed.

vi.  Add fresh media every 10 days or as needed.

vii.  Count the spheres.

viii.  When the spheres have reached a decent size without being dense in the center, dissociate them.

### 11.3.8.2 Dissociation and serial passaging

i.  Carefully collect the media from the wells containing spheres and collect in a 1.5 ml tube.

ii.  Spin down at 300 *g* for 2 minutes and discard most of the supernatant while leaving behind a small amount.

iii. Add 200 μl of acutase to the tubes, vortex, and place in the incubator for 7 minutes.

iv. Pipette up and down vigorously to mechanically break apart the sphere.

v. Add fresh media to the tube and re-plate approximately 1 cell/well into a new 96-well plate (F1 generation).

vi. Incubate the plate at 37 and repeat steps v through viii from the previous section.

vii. When spheres from the F1 plate are observed, repeat the dissociation steps to generate F2 and F3 spheres.

*Chemotherapy selection of CSCs from heterogeneous populations:*

### 11.3.9 Treatment of BCCs with carboplatin to select for CSCs

i. Seed cells at 80%–90% confluency and incubate overnight.

ii. Treat cells every 2–3 days with 200 μg/ml of carboplatin for 2 weeks.

*In vivo serial dilution of CSCs:*

### 11.3.10 Injection of Oct4a^high into NSG mice

i. Perform FACS to isolate Oct4a^high cells. Add media to cells and incubate overnight at 37 °C.

ii. After performing general cell culture, count and collect 200 Oct4a^high cells in 1× PBS.

iii. In a 1:1 ratio, mix the Oct4a^high cells resuspended in 1× PBS with Matrigel to have a final volume of 200 μl.

iv. Inject cells into the dorsal flank of 4-week-old BALB/c mice.

v. Monitor tumors daily for a month and measure tumor volume with a caliper. Tumor volume is calculated with the following formula: $V = \pi r^2 h$, where $r^2$ is the radius and "$h$" is the height.

vi. If the tumor reaches 0.5 cm$^3$ prior to completion of the month, perform surgical resection of the mass and sort the Oct4a^high cells derived from the tumor.

vii. Inject Oct4a^high cells in a similar manner as steps ii–iv for three times.

viii. Perform steps v–vi.

## 11.4 Notes

Note 1: The dose of antibiotic needed for positive selection of Oct4-GFP or SORE6-GFP positive clones will depend on the cell line. A dose-response experiment is recommended to standardize the amount of antibiotic needed for the selection of positive clones from the cell line of interest.

Note 2: Please be aware that every time a virus is thawed, its potency decreases by 10-fold. For this reason, we recommend concentrating the viral stock before storage (refer to Section 11.3.4).

Note 3: NSG mice are immunodeficient, lacking T-cells, B-cells, macrophages, and NK cells. Upon arrival at the institutional animal facility, allow the mice to adapt to the new environment for one week.

## References

[1] Reya T, Morrison SJ, Clarke MF and Weissman IL. Stem cells, cancer, and cancer stem cells. Nature. 2001; 414(6859):105–111.

[2] Kleffel S and Schatton T. Tumor dormancy and cancer stem cells: two sides of the same coin? Adv Exp Med Biol. 2013; 734:145–179.

[3] Dawood S, Austin L and Cristofanilli M. Cancer stem cells: implications for cancer therapy. Oncology. 2014; 28:1101–1107, 1110.

[4] Prieto-Vila M, Takahashi RU, Usuba W, Kohama I and Ochiya T. Drug Resistance Driven by Cancer Stem Cells and Their Niche. Int J Mol Sci. 2017; 18:2574.

[5] Yu Z, Pestell TG, Lisanti MP and Pestell RG. Cancer stem cells. Int J Biochem Cell Biol. 2012; 44(12):2144–2151.

[6] Al-Hajj M, Wicha MS, Benito-Hernandez A, Morrison SJ and Clarke MF. Prospective identification of tumorigenic breast cancer cells. Proc Natl Acad Sci U S A. 2003; 100(7):3983–3988.

[7] Ginestier C, Hur MH, Charafe-Jauffret E, Monville F, Dutcher J, Brown M, Jacquemier J, Viens P, Kleer CG, Liu S, Schott A, Hayes D, Birnbaum D, et al. ALDH1 is a marker of normal and malignant human mammary stem cells and a predictor of poor clinical outcome. Cell Stem Cell. 2007; 1:555–567.

[8] Glumac PM and LeBeau AM. The role of CD133 in cancer: a concise review. Clin Transl Med. 2018; 7:18.

[9] Ahmed SI, Javed G, Laghari AA, Bareeqa SB, Farrukh S, Zahid S, Samar SS and Aziz K. CD133 Expression in Glioblastoma Multiforme: A Literature Review. Cureus. 2018; 10:e3439.

[10] Rahman M, Jamil HM, Akhtar N, Rahman KMT, Islam R and Asaduzzaman SM. Stem cell and cancer stem cell: A tale of two cells. Prog Stem Cell. 2016; 3:97.

[11] Holmberg Olausson K, Maire CL, Haidar S, Ling J, Learner E, Nister M and Ligon KL. Prominin-1 (CD133) defines both stem and non-stem cell populations in CNS development and gliomas. PLoS One. 2014; 9:e106694.

[12] Dontu G, Abdallah WM, Foley JM, Jackson KW, Clarke MF, Kawamura MJ and Wicha MS. In vitro propagation and transcriptional profiling of human mammary stem/progenitor cells. Genes Dev. 2003; 17:1253–1270.

[13] Patel SA, Ramkissoon SH, Bryan M, Pliner LF, Dontu G, Patel PS, Amiri S, Pine SR and Rameshwar P. Delineation of breast cancer cell hierarchy identifies the subset responsible for dormancy. Sci Rep. 2012; 2:906.

[14] Tang B, Raviv A, Esposito D, Flanders KC, Daniel C, Nghiem BT, Garfield S, Lim L, Mannan P, Robles AI, Smith WI, Jr., Zimmerberg J, Ravin R, et al. A flexible reporter system for direct observation and isolation of cancer stem cells. Stem Cell Reports. 2015; 4:155–169.

[15] Guiro K, Patel SA, Greco SJ, Rameshwar P and Arinzeh TL. Investigating breast cancer cell behavior using tissue engineering scaffolds. PLoS One. 2015; 10:e0118724.

# 12

# Bone Marrow Aspirate for Isolation of Mesenchymal Stem Cells

**Lauren S. Sherman, Andrew Petryna, and Joshua Kra**

**Corresponding Author:** Joshua Kra, Department of Medicine, Division of Hematology/Oncology, Rutgers New Jersey Medical School, Rutgers Cancer Institute of New Jersey at University Hospital, USA
E-mail: jk1393@cinj.rutgers.edu
**Disclaimer:** The authors have no conflict to declare.

## Abstract

The adult healthy bone marrow is the source of hematopoietic and non-hematopoietic cells. The latter includes cells such as mesenchymal stem cells (MSCs) and endothelial progenitors. Hematopoietic cells include stem cells (HSCs) and progenitors of various lineages. Studies of bone marrow cells can provide information about various diseases. Thus, it might be required to isolate several primary cells, immediately after harvesting from the bone marrow or by expansion in culture. In order to do this, it is required to first obtain bone marrow aspirate from human donors. This protocol describes a stepwise procedure to collect aspirates from human donors. This collection is required to complete the methods in other chapters, including culture of human MSCs and isolation of HSCs.

## 12.1 Introduction

Bone marrow aspiration is a clinical procedure to analyze the bone marrow for diagnostic purposes. The earliest precursor of bone marrow aspiration is considered to be trepanning of the skull bone, which is considered one of the oldest medical procedures to be performed by man [1, 2]. Since first noted in skeletons from the Neolithic and pre-Columbian Andean civilizations 8000–10,000 years ago, the tools and method of trepanning – or driving a hole through a bone – advanced with other developments in man [3]. Trepanning continued to be a common treatment for head injuries until the early 1800s when surgeries transitioned from the home to the hospital setting, at which point trepanning fell out of favor due to the high rates of infection [1]. As science progressed with the development of antiseptics and infection-prevention methods in the late 1800s, trepanation was again widely used to treat head trauma [1].

Trepanning was proposed as a method to study the bone marrow in the late 1800s, and in the early 1900s, the sternum became the target bone of choice for *in vivo* study of the bone marrow [1]. Similarly, fluid resuscitation via sternum injection was described for treating hypovolemic shock in World War II – although this method fell out of favor with the introduction of plastic catheters for intravenous access [1]. In 1950, the iliac crest was identified as a rich source of bone marrow, opening the opportunity of studying the marrow in a more accessible, safer site [4]. Over the next decade, the technique was optimized, identifying the posterior iliac crest in particular as an optimal site and improving specific trepanation tools and techniques, ultimately leading to the current technique [1].

Bone marrow aspiration remains the common method in clinical diagnosis for hematological disorders such as anemia and leukemia. Aspirates can be used for other reasons such as selection of different hematopoietic cell subsets. This method describes the clinical technique used to collect the bone marrow aspirate.

## 12.2  Materials

All items should be sterile and clinical grade where appropriate.

1.  Institutional Review Board Approval with informed consent

2.  Pen or marker

3.  Surgical gloves

4.  Sterile field: either a procedure drape, several large pieces of sterile sponge, sterile towels, or similar

5.  Betadine solution (10% povidone iodine), chlorhexidine gluconate, or alternative topical antiseptic

6.  Sterile all-purpose sponges/gauze, 4 in × 4 in

7.  10 cc syringes, divided

8.  Heparin solution, 1000 U/ml

9.  21-Gauge (G), 1/2 inch hypodermic needles

10. 25G, 5/8 inch hypodermic needles

11. Alcohol sterilization pads

12. Lidocaine HCl, 1%–2%

13. Bone marrow aspiration needle, 15 GA × 4 in (Argon Medical Devices Ref DBMNI1504)

14. Adhesive bandage

15. Biospecimen bag

16. Optional: saline (flush syringe)

## 12.3 Methods

### 12.3.1 Human donor

It is generally preferable to collect bone marrow aspirates from generally healthy donors between 18 and 30 years old. However, this will depend on the research. As an example, for research involving an aging study, researchers might want to select a donor, >60 years or whatever the scientists consider as aged. It should be noted that when isolating MSCs from older donors, the MSCs tend to passage less efficiently as compared to MSCs from younger donors. Unless the research requires samples from a particular subject population, it is recommended that the bone marrow donor be selected only if healthy – no infection or taking medication. These risks might alter cell behavior, adding confounding factors and artifact *in vitro*. From a clinical perspective, the donor should be screened for anticoagulant use as this could cause unwanted bleeding.

Once a donor is identified, informed consent must be obtained prior to taking the aspirate. Informed consent should include the possible risk (e.g., infection, bleeding, or pain) that the donor is under no obligation to donate bone marrow and that results of this study are unlikely to directly affect them (unless this statement is not applicable to a particular project).

Donors can repeat the bone marrow donation process after 3–6 months after the previous donation. The hematologist who is taking the aspirate, in consultation with the donor, will determine if the procedure should be taken from a different side of the iliac crest.

## 12.3.2 Bone marrow aspiration

This is a mildly invasive procedure, generally performed by a hematologist.

### 12.3.2.1 Preparation

1.  Prior to obtaining bone marrow aspirate, if the intent is to isolate MSCs, prepare the tissue culture plates (see Chapter 3).

2.  Prepare five 10 cc syringes with 1 ml heparin solution at 1000 units/ml (Notes 1 and 2). Replace the syringes containing heparin to the original packages for transport to the clinic. It is recommended that the syringes be placed in biospecimen bags for transport.

3.  Prepare a procedure tray with all necessary materials; see materials listed above.

4.  The donor should be asked to expose the area around the iliac crest (e.g., loosen their pants). This will be followed by the donor lying in a lateral decubitus position.

5.  Palpate the iliac crest region to identify a preferred site for aspiration. The site should be marked with a pen or marker (Note 3).

6.  Cleanse the procedure field with a betadine-soaked sponge (Note 4).

7.  Apply the sterile field surrounding the aspiration site. At this time, the hematologist should be wearing sterile surgical gloves.

### 12.3.2.2 Anesthetizing the region

1.  Prepare the lidocaine filled syringe. If using a multi-dose vial, cleanse the rubber stopper with an alcohol pad. Using a 25G needle attached to a 10 cc syringe, load approximately 10 ml of lidocaine.

2.  Anesthetize the skin and underlying tissues with approximately 0.5−1 ml of lidocaine.

3.  Remove the needle and recap the lidocaine-filled syringe with a 21G needle. The 21G needle is then used to anesthetize the subcutaneous tissue and the underlying periosteum.

4.  Ensure adequacy of anesthesia by gently tapping the periosteum with the needle tip, asking the donor if they feel any painful sensation(s). The donor may feel pressure but should not feel pain. If the donor reports

pain, additional lidocaine should be administered. Once the area is sufficiently anesthetized, remove the needle (Note 5).

### 12.3.2.3 Aspirating the marrow

1. Insert the bone marrow aspiration needle at the marked site. Once the bone is reached, advance the needle slowly, alternately rotating the needle clockwise and counterclockwise until the needle penetrates the cortical bone and enters the marrow cavity (Note 6).

2. Remove the internal plug from the bone marrow aspiration needle and connect a heparin-containing 10 cc syringe to the aspiration needle. Collect a small volume of marrow to confirm that the aspiration needle is within the marrow cavity. If a dry tap is produced, reinsert the plug and advance slightly farther into the marrow cavity.

3. Aspirate 3–6 ml of marrow into each heparin-loaded syringe. It is not recommended to aspirate >20 ml as, at this point, the aspirate will predominantly be peripheral blood rather than marrow (Note 7).

4. Return the marrow-filled syringes to the wrappers and/or a biospecimen bag and immediately invert the syringe several times to mix the marrow with the heparin solution to prevent coagulation.

5. Transport the collected bone marrow aspirate to the laboratory for processing.

### 12.3.2.4 Post-procedural care

1. Reinsert the bone marrow aspiration needle's internal plug and remove the marrow needle.

2. Apply pressure to the aspiration site with gauze until any bleeding has stopped. In healthy donors, this will usually take place within 2–5 minutes; however, it may take longer.

3. Wipe off residual betadine with saline and additional gauze.

4. Apply an adhesive bandage above the aspiration site. Then ask the donor to sit up and apply pressure to the site for 5 minutes.

5. Instruct the donor to keep the site dry for 24 hour and to check regularly during this time to ensure that bleeding does not recur. If minor bleeding does occur, pressure should be applied until the bleeding stops and

a fresh bandage applied. If bleeding persists, the donor should return to the clinic.

## 12.4 Notes

Note 1: To load the syringe with heparin, prepare stock solution under aseptic conditions (i.e., within a tissue culture hood). Load the syringe using a 21G or larger hypodermic needle.

Note 2: 1000 units of heparin is sufficient for 5–6 ml of bone marrow aspirate. If the hematologist opted for a different volume of aspirate, adjust the volume of heparin accordingly.

Note 3: If there is any difficulty identifying the posterior iliac crest, ask the donor to place their hand on the hip: this will facilitate identification of the pelvic rim as compared to the lateral sacral crest. Misidentification of the lateral sacral crest will lead to a dry tap.

Note 4: Instead of povidone iodine, you may use any clinical topical antiseptic.

Note 5: The syringe used for anesthesia can be left in its anatomic site while preparing the aspiration needle. This will help ensure that the same location is used for the bone marrow aspiration.

Note 6: A sudden reduction in pressure will indicate that the marrow cavity has been entered. Do not penetrate too deep into the marrow.

Note 7: It is often beneficial to distract the patient through breathing and conversation. Doing so will help the patient relax and reduce anxiety.

## References

[1] Hernigou P. The history of bone marrow in orthopaedic surgery (part I trauma): trepanning, bone marrow injection in damage control resuscitation, and bone marrow aspiration to heal fractures. Int Orthop. 2020; 44(4):795–808.
[2] Parapia LA. Trepanning or trephines: a history of bone marrow biopsy. Br J Haematol. 2007; 139(1):14–19.
[3] Rifkinson-Mann S. Cranial surgery in ancient Peru. Neurosurgery. 1988; 23(4):411–416.
[4] Rubinstein MA. The technic and diagnostic value of aspiration of bone marrow from the iliac crest. Ann Intern Med. 1950; 32(6):1095–1098.

# 13

# Culturing Human Bone Marrow Stromal Cells

Yannick Kenfack[1,2], Lauren S. Sherman[1,2], Bobak Shadpoor[1,2], Andrew Petryna[1,2], Sami Souyah[1], Ella Einstein[1], Stephanie Perricho[1], and Pranela Rameshwar[1]

[1]Division of Hematology/Oncology, Department of Medicine, New Jersey Medical School, Rutgers Biomedical and Health Sciences, USA
[2]Rutgers School of Graduate Studies at New Jersey Medical School, USA
**Corresponding Author:** Pranela Rameshwar, Rutgers New Jersey Medical School, MSB, USA
E-mail: rameshwa@njms.rutgers.edu
**Disclaimer**: The authors have nothing to declare.

## Abstract

Bone marrow (BM) stromal cells (BMSC) comprise hematopoietic and non-hematopoietic cells, such as macrophages, fibroblasts, adipocytes, and endothelial cells. These cells are considered to be hematopoietic supporting cells and have been used to assess functions within the BM microenvironment using *in vitro* models. These cells are critical when evaluating the functions of hematopoietic stem cells *in vitro* using the long-term culture initiating assay. Unlike solid organs that can be easily removed for research purposes, the BM cannot be similarly harvested. Thus, to understand bone marrow functions, model systems require the hematopoietic stem and progenitor cells to be studied within the endogenous microenvironment. The collection of cells and factors within the stromal culture represent this niche environment. In this chapter, we present a step-by-step protocol for isolating stromal cultures from bone marrow aspirates.

## 13.1 Introduction

The bone marrow cavity is home to hematopoietic and non-hematopoietic cells [1]. These include the supporting stromal cells, hematopoietic stem cells (HSCs), progenitors, mesenchymal stem cells, and immune cells. HSCs are the source of life-long hematopoiesis that produce blood and immune cells [2]. Aging of the hematopoietic system can lead to several disorders such as myeloid and lymphoid malignancies [3]. The disorder of the BM is not only limited to hematological malignancies but also to solid tumors. In fact, the marrow is the preferred organ for breast cancer cells [4]. Breast cancer cells can take advantage of the BMSCs to survive as dormant cells for decades [5]. In addition, BMSC can also support B-cell malignancies [6]. Given the key roles of BMSCs in homeostasis and malignant BM functions, it is important to include this chapter that describes a step-by-step procedure to culture these cells.

BMSCs when *in vitro* derived a layer of mostly non-hematopoietic cells that can interact with HSCs and secrete extracellular matrices [7]. The cells are mostly fibroblasts with relatively less endothelial cells, reticular cells, adipocytes, and M2 macrophages [8]. Culture of BMSCs allow for scientists to establish a supporting layer of cells to study the functions of HSCs in a system referred to as long-term culture initiating cells [9]. We describe this method in detail beginning with bone marrow aspirates (Chapter 12). The method was adapted from ref. [10].

## 13.2 Materials and Reagents

### 13.2.1 BMSC tissue culture media

1.  α-Minimum essential medium (α-MEM) (ThermoFisher Scientific)

2.  Fetal bovine serum (FBS) (Gibco – ThermoFisher Scientific)

3.  Penicillin–streptomycin (P-S) (Gemini Bioproducts)

4.  L-Glutamine (Gibco – ThermoFisher Scientific)

5.  Horse sera (HS), Hyclone

    - Inactivate heat at 56 °C for 45 minutes and then aliquot into 100 ml volumes in sterile bottles. Freeze at − 22 °C. Thaw as needed. Leave the thawed bottle at 4 °C as the working HS.

- β-Mercaptoethanol (β–ME) (Bio-Rad), 14.3 M.

  a. 0.2 ml of stock (14.3 M) β-ME in 4.8 ml tissue culture grade PBS, pH 7.4

     $= 5 \times 10^{-1}$ M β–ME

  b. 1 ml of "a" in 4 ml PBS

  c. 1 ml of "b" in 9 ml PBS

     $= 1 \times 10^{-2}$ M β–ME (Filter Sterilize); store at $4\,°C$

6. Hydrocortisone 21-sodium succinate, m.w. =484.50

   - Prepare fresh solution every week: 0.4845 mg/10 ml -MEM; Filter sterilize; store at $4\,°C$.

| Reagent | Final concentration | Volume |
|---|---|---|
| Horse sera | 12% | 12.5 ml |
| FCS | 12% | 12.5 ml |
| α-MEM with P-S | | 72.9 ml |
| β-ME | $10^{-4}$M | 1 ml |
| Glutamine | 1.6 mM | 1 ml |
| Hydrocortisone | $10^{-7}$ M | 0.1 ml |

### 13.2.2 Complete BMSC media

BM aspirate:

1. See Chapter 12 for the method to acquire human marrow aspirate.

## 13.3 Method

1. Allow BM samples to stand at room temperature until there is an obvious layer of plasma, buffy coat, and packed cells.

2. Remove buffy coat and resuspend in stroma medium.

3. Seed $10 \times 10^6$/ml in T25 flasks.

4. Incubate for four days at $37\,°C$, 5% $CO_2$.

5. Remove red cells and granulocytes by Ficoll-Hypaque density gradient. Wash mononuclear cells and add back to the flasks.

6. Feed weekly by replacing half of the culture medium with fresh medium. Stroma should be confluent by week 3.

7. Wash thoroughly with PBS or any media to remove non-adherent cells before using the adherent stroma cells for experiments.

8. To passage all stroma subsets, treat the adherent cell with 0.2% collagenase or trypsin, count, and then subculture.

## References

[1] Omatsu Y. Cellular niches for hematopoietic stem cells in bone marrow under normal and malignant conditions. Inflamm Regen. 2023; 43(1):15.

[2] Brown G. Lessons to cancer from studies of leukemia and hematopoiesis. Front Cell Dev Biol. 2022; 10:993915.

[3] Xie X, Su M, Ren K, Ma X, Lv Z, Li Z, Mei Y and Ji P. Clonal hematopoiesis and bone marrow inflammation. Transl Res. 2022.

[4] Yang R, Jia L, Lu G, Lv Z and Cui J. Symptomatic bone marrow metastases in breast cancer: A retrospective cohort study. Front Oncol. 2022; 12:1042773.

[5] Lim PK, Bliss SA, Patel SA, Taborga M, Dave MA, Gregory LA, Greco SJ, Bryan M, Patel PS and Rameshwar P. Gap junction-mediated import of microRNA from bone marrow stromal cells can elicit cell cycle quiescence in breast cancer cells. Cancer Res. 2011; 71(5):1550–1560.

[6] Mangolini M and Ringshausen I. Bone Marrow Stromal Cells Drive Key Hallmarks of B Cell Malignancies. Int J Mol Sci. 2020; 21(4).

[7] Dorshkind K. Regulation of Hemopoiesis by Bone Marrow Stromal Cells and Their Products. Annu Rev Immunol. 1990; 8(1):111–137.

[8] Walker ND, Elias M, Guiro K, Bhatia R, Greco SJ, Bryan M, Gergues M, Sandiford OA, Ponzio NM, Leibovich SJ and Rameshwar P. Exosomes from differentially activated macrophages influence dormancy or resurgence of breast cancer cells within bone marrow stroma. Cell Death & Disease. 2019; 10(2):59.

[9] Deryugina EI and Müller-Sieburg CE. Stromal cells in long-term cultures: keys to the elucidation of hematopoietic development? Crit Rev Immunol. 1993; 13(2):115–150.

[10] Hannocks M-J, Oliver L, Gabrilove JL and Wilson EL. Regulation of Proteolytic Activity in Human Bone Marrow Stromal Cells by Basic Fibroblast Growth Factor, Interleukin-1, and Transforming Growth Factor β. Blood. 1992; 79(5):1178–1184.

# 14

# Isolating Mononuclear Cells by Ficoll-Hypaque Density Gradient

**Andrew Petryna[1,2], Lauren S. Sherman[1,2], Anushka Sarkar[1,2], Bobak Shadpoor[1,2], Yannick Kenfack[1,2], and Pranela Rameshwar[1]**

[1]Division of Hematology/Oncology, Department of Medicine, Rutgers New Jersey Medical School, USA
[2]Rutgers School of Graduate Studies at New Jersey Medical School, USA
**Corresponding Author:** Pranela Rameshwar, Department of Medicine – Division of Hematology/Oncology, Rutgers New Jersey Medical School, USA
E-mail: rameshwa@njms.rutgers.edu
**Disclaimer:** The authors have nothing to declare.

## Abstract

In cases when different stem cells are derived from tissues such as bone marrow aspirate, umbilical cord blood, and mobilized peripheral blood, it is important to eliminate red blood cells and neutrophils. This method is effective by density gradient such as Ficoll-Hypaque. The density will vary depending on the species and this should be noted at the beginning of the method. In the case of isolation of mesenchymal stem cells (MSCs) or stromal cells from the bone marrow, it is critical after three days to eliminate red blood cells (RBCs) by density gradient. This will remove the RBCs before they are lysed, which could be toxic to the adherent cells. Additionally, the isolation of hematopoietic cell subsets from human marrow aspirate would require starting with mononuclear, which would be derived from density gradient. This method describes a step-by-step process to isolate mononuclear cells by density gradient.

## 14.1 Introduction

Over the last two decades, much interest has been generated on the potential uses of mesenchymal stem cells (MSCs) in research and clinical trials. Reproducible data from various research groups across the globe require culture of large number of cells. Similarly, the hematopoietic system requires studying various cell subsets from tissues such as bone marrow aspirates, umbilical cord blood, and mobilized peripheral blood. Bone marrow contains a mix of arterial and venous blood, and, therefore, aspirates will comprise multiple hematopoietic cells as well as red blood cells [1].

The major issue when studying the aforementioned tissue products is the presence of red blood cells. This is particularly relevant when culturing adherent cells from bone marrow aspirate, e.g., MSCs (Chapter 3) and stromal cultures (Chapter 13). As described in these methods, after day 3, the red blood cells (RBCs) will begin to lyse, and, if not removed, the lysates will be toxic to the desired adherent cells. In this regard, the non-adherent cells are generally subjected to the Ficoll-Hypaque density gradient to eliminate the RBCs and neutrophils and the mononuclear fraction can be replaced in the culture.

Ficoll-Hypaque is an inert polysaccharide of high molecular weight, which is generally used to separate cell populations of varying densities [2]. Using a preparation of Ficoll compatible with human blood cell densities, a layered gradient can be created to separate mononuclear cells from RBCs and neutrophils with the starting tissues, that is, blood or bone marrow aspirate [3]. In cases where it is desired to retain plasma for the collection of blood products such as microvesicles, the Ficoll method allows for simultaneous collection of mononuclear cells and plasma. In general, this chapter supplements several other methods to expand other stem cells.

## 14.2 Materials

All items should be tissue culture grade or in cases where a labware is reusable, sterilize by autoclaving (Note 1).

1. Ficoll-Histopaque 1077 for separation of human mononuclear cells (Note 2)

2. Pasteur pipette with cotton plugs

3. Tissue culture grade sterile phosphate buffered saline, pH 7.4 (PBS)

4. 15 ml polypropylene or polystyrene sterile conical centrifuge tubes (Note 3)

5. 50 ml polypropylene or polystyrene sterile conical centrifuge tubes (Note 3)

## 14.3 Methods

### 14.3.1 Sample

If you are performing density gradient separation on bone marrow aspirates, refer to Chapter 12. On the other hand, if you are separating mononuclear cells from whole peripheral blood, the amount of starting blood will depend on the desired number of cells. As a rule of thumb, 1 ml of blood from a healthy individual (18–35 years) is expected to yield ~3–5 × $10^6$ mononuclear cells (Note 4).

### 14.3.2 Loading onto Ficoll-Hypaque (Note 2)

1. If you need the cells to be sterile, perform the process in a laminar flow tissue culture hood.

2. Add Ficoll-Histopaque 1077 equal to the amount of blood product in a conical tube. If you have ~10 ml of blood, the final volume with Ficoll would be ~20 ml. In this case, use a 50 ml centrifuge tube. You should always add the Ficoll to the tube before beginning to add the blood.

3. Use a sterile Pasteur pipette or a regular sterile pipette to carefully transfer the blood on top of the Ficoll. Place the tip of the pipette containing the blood against the inside of the tube, at least a centimeter above the Ficoll surface. Slowly add the blood to prevent mixing with Ficoll (Note 5). Avoid using a serological pipette larger than 5 ml. After adding the samples, you should notice a distinct separation between the layer of blood and Ficoll base.

   - Instead of adding the blood with the tube in a vertical position, you might want to use a centrifuge tube to gain an angle between 45° and 80° before adding the blood. Proceed to add the blood as above (Note 6).

4. Immediately centrifuge the tubes at 500 *g* for 30 minutes at room temperature. Turn the brake to an off position (Notes 7 and 8).

5. After centrifugation, immediately remove the tubes and transfer to the tissue culture hood.

6.  If you are separating blood or bone marrow aspirates and you need the top layer of plasma, collect and add to a clean tube. Be careful not to disturb the mononuclear layer. If you are not interested in the plasma, carefully aspirate the top layer. You should leave a small amount of the top layer since this will avoid aspiration of the valuable mononuclear fraction.

7.  Carefully collect the mononuclear layer, referred as buffy coat with a Pasteur pipette. This is the cloudy but distinct layer above the Ficoll (Note 9). Transfer the buffy coat to a conical centrifuge tube. Although you should collect some of the Ficoll to ensure acquiring all the cells, be careful to avoid collecting a relatively large amount of Ficoll. The latter is dense and will add to the density of the washing solution. This will provide a challenge to efficiently pellet the cells (next steps) (Note 10).

8.  Add sera-free media to fill the tube (Note 11). This will dilute the residual Ficoll and plasma to decrease the density. Centrifuge the tube at 500 $g$ for 10 minutes. You may turn on the brake at this time. If you believe that you have added too much Ficoll, increase the centrifugation time. You should also examine the tube after centrifugation to determine if the supernatant is cloudy. The latter will indicate that there are substantial cells remaining in the supernatant. If so, continue to centrifuge the tubes until the supernatant is clear. Select the centrifugation time based on the cloudiness of the supernatant. This first centrifugation is not considered a wash. If you desire to wash the cells, move to the next step.

9.  Immediately after you have removed the tube from the centrifuge, move to a laminar flow hood. Aspirate the supernatant without touching the pellets. To avoid aspirating the cells, leave a small amount of media on top of the pellet.

10. Use your fingers to tap the end of the tube to resuspend the pellet (Note 12).

11. In cases when you are isolating the mononuclear fraction to eliminate red blood cells (RBCs) as for stroma or mesenchymal stem cell cultures, replace the mononuclear cells within the buffy coat to the respective cultures (Methods 3 and 13). In these instances, you do not need to wash the mononuclear cells.

12. Washing: Fill the tube with media or tissue culture grade PBS and centrifuge for 10 minutes if a 15 ml tube or 20–30 minutes for 50 ml tubes. Repeat once more.

13. Once you are satisfied with your washes and have your final pellet, you may resuspend your pellet in the appropriate complete media, depending on the desired assay or cell expansion.

## 14.4 Notes

Note 1: Examples of autoclavable products could be bottles for aliquoting of media.

Note 2: The described method is described for human samples. If you are separating cells from animal blood or related tissues, refer to the commercial catalogue for the appropriate Ficoll density. Ficoll-Histopaque 1077 is a density gradient buffer solution. The 1077 denotes a density of 1.077 g/ml, optimized for the separation of human blood components. Other density solutions made for this purpose will have different numbers succeeding them to denote their use for specific species.

Note 3: If you plan to retain the plasma for quantification of cytokines, it is better to use polypropylene tubes. This will prevent absorption onto polystyrene surface.

Note 4: If you have limited samples, as would be expected with patient samples, you could dilute the blood with tissue culture media. This will allow you to recover the maximum number of mononuclear cells.

Note 5: As an alternative to adding the blood on top of the Ficoll, you could place the pipette containing the blood at the bottom of the tube and then slowly release the sample.

Note 6: If you allow the sample to mix with the Ficoll instead of gently layering on top of it, you will not get a clear separation. You could try to place the tube at an angle and use a small pipette. This will minimize the strength of the stream of blood being released and prevent the breaking of the Ficoll's surface tension.

Note 7: If you allow the tubes to sit for a prolonged period after adding the top layer, the cells will begin to enter the Ficoll. This will hinder proper isolation of the mononuclear fraction. Also, Ficoll could be toxic to the cells. Thus, avoid prolonged exposure.

Note 8: Try to accelerate and decelerate slowly to prevent mixing of the different layers. Understand how your centrifuge applies brakes to throttle acceleration and deceleration. The rate of change of speed must be as gradual as possible to prevent disturbing of the layers.

Note 9: If the blood is from a patient with hematological disorder, you are likely to have some mixture between the buffy coat (mononuclear cells) within the Ficoll and plasma. Specifically, the clean separation of the mononuclear cells that is expected with healthy blood might not be the same with patients' samples.

Note 10: Following centrifugation, there are three layers: the top layer containing plasma, a thin cloudy layer of mononuclear cells called the buffy coat, the Ficoll, and a pellet of red blood cells at the very bottom. It is the Buffy coat where the MSCs and PBMCs are found. It is often quite thin; so you must take care to avoid eliminating this layer when you are aspirating the top layer of plasma.

Note 11: In general, human mononuclear cells can survive without sera or media, which will contain nutrients. If you are not worried about this and you would like to have less cost, use PBS instead.

Note 12: If the cells are not immediately dislodged from the pellet, you are likely to have low viability.

## References

[1] Harrison JS, Rameshwar P, Chang V and Bandari P. Oxygen saturation in the bone marrow of healthy volunteers. Blood. 2002; 99(1):394–394.
[2] Noble P and Cutts J. Separation of blood leukocytes by Ficoll gradient. Canadian Vet J. 1967; 8(5):110.
[3] Chang Y, Hsieh P-H and Chao C. The efficiency of Percoll and Ficoll density gradient media in the isolation of marrow derived human mesenchymal stem cells with osteogenic potential. Chang Gung Med J. 2009; 32(3):264–275.

# 15

# Phenotypic and Multipotent Characterization of Bone-Marrow-derived Mesenchymal Stem Cells

**Anuska Sarkar[1,2], Andrew Petryna[1,2], Bobak Shadpoor, Lauren S. Sherman[1,2], Pranela Rameshwar[1]**

[1]Division of Hematology/Oncology, Department of Medicine, Rutgers New Jersey Medical School, USA
[2]Rutgers School of Graduate Studies at New Jersey Medical School, USA
**Corresponding Author:** Pranela Rameshwar, Department of Medicine – Division of Hematology/Oncology, Rutgers New Jersey Medical School, USA
E-mail: rameshwa@njms.rutgers.edu
**Disclaimer:** The authors have nothing to declare.

## Abstract

Mesenchymal stem cells (MSCs) are non-hematopoietic multipotent cells. MSCs are ubiquitous with high frequency in adipose tissue, dental pulp, umbilical cord, placenta, and bone marrow. Their therapeutic potential, along with reduced ethical concern, and ease of isolation and expansion contribute to the interest of these categories of stem cells in translational research. This chapter adds to another section that describes the method to culture MSCs. Culturing of MSCs, regardless of the source, should have a basic method to ensure the integrity of the cultured cells. This chapter selects MSCs from the bone marrow to demonstrate quality control that could be used to characterize similar cells from all sources. Phenotypic quality control is assessed by flow cytometry for specific surface markers and multilineage differentiation, tri-lineage differentiation into adipocytes, chondrocytes, and osteocytes. MSCs are generally thought to be mesodermal. Thus, investigators could examine transdifferentiation to neurons, depending on the expertise.

## 15.1 Introduction

Mesenchymal stem cells (MSCs) are non-hematopoietic adult stem cells that can differentiate into cells of any germ layer [1]. MSCs have been well-studied with respect to generate cartilage, ligaments, bone, and connective tissue [2]. The emergence of MSCs as candidates for stem cell therapies has led to the development and optimization of various techniques for their isolation, ex vivo expansion, and characterization.

The International Society for Cellular Therapy (ISCT) provided guidance to characterize MSCs using specific surface markers such as CD44, CD73, CD90, and CD105 and the absence of hematopoietic markers [3]. Any laboratory will need to implement quality control as they expand MSCs. This is easily accomplished by flow cytometry for surface markers. In this method, cells flow through a laser beam in a hydrodynamically focused fluid stream. The difference in the scattering of light and fluorescence emitted by the immunolabeled cells are used to determine cellular properties like size, internal complexity, and surface antigen expression can be measured on [4].

Flow cytometry can assess the expected phenotype of cultured MSCs. However, this sole assessment is inadequate and should include multipotency, which could be assessed by multilineage differentiation [5]. This can be achieved by examining the ability of the MSCs to differentiate into cells of mesodermal lineages – adipocytes, chondrocytes, and osteocytes using well-described methods [6].

MSCs can transdifferentiate into cells of neuronal lineage and could facilitate research on conditions associated with the nervous system [7]. Several methods have been reported to induce MSCs into neuronal cells. A true multipotent MSC would be able to form cells of various germ layers. Thus, examination of the cultured MSCs can form neuronal-like cells and could be an advantage when assessing the function of MSCs. This could be achieved with specific media to induce neuronal transdifferentiation, which would express proteins such as NeuN, Tau, and Nestin [8–10].

Since this volume has several chapters on the isolation of MSCs from different sources, it is necessary to describe how these expanded cells are characterized. We select MSCs expanded from human bone marrow aspirate as an example to describe the basic quality control to assess the cells' integrity. We describe phenotype by flow cytometry while discussing the advantage of testing the following markers: CD44, CD73, CD90, and CD105. We also indicate the importance of ensuring the lack of hematopoietic markers, such as CD11b, CD19, CD34, and CD45 [5]. MHC-II can be expressed on MSCs, which is consistent with these cells being antigen presenting cells and negative MHC-II to correlate when licensed as immune suppressor cells within an inflammatory microenvironment [11, 12]. We also describe the method to assess multipotency by tri-lineage into mesodermal lineages and ectodermal differentiated cells [13].

## 15.2 Materials and Reagents

### 15.2.1 Flow cytometry

#### 15.2.1.1 Reagents

i.  Trypsin/EDTA solution (Gibco, Cat No. 25200056) (Note 1)

ii. Labeling buffer (BD Cat No. 554656) or self-prepared solution (Note 2):

   a. PBS, pH 7.4 (this does not have to be tissue culture grade)

   b. 2%–5% FBS (R&D Systems Cat No. S11150)

   c. 0.09% Sodium azide (Sigma Cat No. S2002)

iii. 1 × PBS (phosphate buffered saline solution)

iv. Human MSC Analysis Kit (BD Stemflow™, Cat No. 562245) or component conjugated antibodies (Note 3):

   a. hMSC positive cocktail antibodies (Note 4):

   - BD Pharmingen™ APC Mouse Anti-human CD105 Clone 266 (Cat No. 562408)

   - BD Pharmingen™ PE Mouse Anti-human CD90, Clone 5E10 (Cat No. 555596)

   - BD Pharmingen™ PerCP-Cy™ 5.5 Mouse Anti-human CD73, Clone AD2 (Cat No. 561260)

   b. Human MSC (hMSC) negative cocktail antibodies (Note 5):

   - BD Pharmingen™ FITC Mouse Anti-human CD34, Clone:581 (Cat No. 555821)

   - BD Pharmingen™ FITC Mouse Anti-human CD11b, Clone: ICRF44 (Cat No. 562793)

   - BD Pharmingen™ FITC Mouse Anti-human CD14, Clone: M5E2 (Cat No. 555397)

- BD Pharmingen™ FITC Mouse Anti-human CD45RA, Clone: HI100 (Cat No. 555488)

- BD Pharmingen™ FITC Mouse Anti-human HLA-DR, Clone: G46-6 (Cat No. 555811)

   c.  Isotype Controls (Note 6):

- BD Pharmingen™ FITC Mouse IgM, κ Isotype Control, Clone: G155-228 (Cat No. 555583)

- BD Pharmingen™ FITC Mouse IgG1, κ Isotype Control, Clone: MOPC-21 (Cat No. 555909)

   d.  Positive hMSC Isotype Control Antibodies (Note 7):

   BD™ Mouse IgG1 PE, Clone: X40 (Cat No. 340761)

   BD™ Mouse IgG1 PerCP-Cy™ 5.5, Clone: X40 (Cat No. 347212)

   BD™ Mouse IgG1 APC, Clone: X40 (Cat No. 340754)

 v.  BD™ LSRII or other similar flow cytometry system

vi.  Thermo Fisher Sorvall ST16R or similar centrifuge system

## 15.2.2 Tri-lineage differentiation

## 15.2.2.1 Reagents for Osteogenic differentiation assay

 i.  MSC Differentiation BulletKit™-Osteogenic (Lonza, Cat No. PT-3002):

   a.  1 × MSC Osteogenic Differentiation Basal Medium (Lonza, Cat No. PT-3924)

   b.  1 × MSC Osteogenic Differentiation SingleQuots™ Supplements Kit (Lonza, Cat No. PT-4120)

   c.  Trypsin/EDTA solution (Lonza, Cat No. CC-3232)

ii.  Hydrochloric acid (HCl)

## 15.2.2.2 Reagents for Chondrogenic differentiation assay

 i.  MSC Differentiation BulletKit™-Chondrogenic (Lonza, Cat No. PT-3003):

   a.  1 × MSC Chondrogenic Basal Medium (Lonza, Cat No. PT-3925)

b. 1 × MSC Chondrogenic SingleQuot Kit (Lonza, Cat No. PT-4121)

c. TGF-β3 (Lonza, Cat No. PT-4124)

ii. 1 mg/ml BSA

### 15.2.2.3 Reagents for Adipogenic differentiation assay

i. MSC Differentiation BulletKit™-Adipogenic (Lonza, Cat No. PT-3004):

a. 1 × hMSC Adipogenic Maintenance Medium (Lonza, Cat No. PT-3102A)

b. 1 × hMSC Adipogenic Induction Medium (Lonza, Cat No. PT-3102B)

c. 1 × hMSC Adipogenic Maintenance SingleQuot Kit (Lonza, Cat No. PT-4122)

d. 1 × hMSC Adipogenic Induction SingleQuot Kit (Lonza, Cat No. PT-4135)

e. AdipoRed™ assay reagent (Lonza, Cat No. PT-7009)

### 15.2.3 Peptidergic neuronal transdifferentiation [13]

### 15.2.3.1 Neuronal induction media

i. Fetal bovine serum (R&D Systems)

ii. Dulbecco's Modified Eagle's Medium, High-glucose, Sodium bicarbonate (Sigma-Aldrich Cat No. D567)

iii. Dulbecco's Modified Eagle's Medium/Nutrient Mixture F-12 Ham (Sigma-Aldrich Cat No. D8062)

iv. All-trans retinoic acid (RA) (Sigma, Cat No. R 2625)

v. Recombinant human basic fibroblast growth factor (bFGF) (R&D Systems Cat No. 3718-FB)

vi. B-27 Neuronal Supplement (Gibco Cat No. 17504044)

### 15.2.3.2 Immunofluorescence

i. 1 × PBS

ii. 3.7% Formaldehyde

iii. 1% Triton-X (Sigma-Aldrich Cat No. X100)

iv. 4,6-Diamidino-2-phenyl indole, dilactate (DAPI) (Molecular Probes/ Invitrogen™, Cat No. D3571)

v. Texas Red™-X Phalloidin (Molecular Probes/Invitrogen™, Cat No. T7471)

vi. Mouse Anti-Neuronal Nuclei (NeuN) mAb (Chemicon, Cat No. MAB377)

vii. Rabbit anti-MAP2 (Chemicon Cat No. AB2290)

viii. Goat anti-Tau (Chemicon, Cat No. AB5868)

ix. Mouse anti-glial fibrillary acidic protein (GFAP) (Chemicon, MAB3402)

x. Phycoerythrin (PE) goat anti-rabbit (abcam Cat No. ab59588)

xi. PE-Rabbit anti-goat (abcam Cat No. ab72465)

xii. Fluorescein isothiocyanate (FITC) goat anti-mouse (abcam Cat No. ab6785)

xiii. FITC-goat anti-rabbit (abcam Cat No. ab6717)

xiv. Three-color fluorescent microscope (Nikon Instruments Inc.)

## 15.3 Methods

This method is limited to characterization. Refer to other chapters in this volume for methods to culture MSCs from different sources.

### 15.3.1 Flow cytometry

Laboratories can purchase kits with antibodies could proceed. However, for laboratories with limited resource, specific antibodies can be purchased. Three antibodies can be used, along with CD45 as negative control and iso-type control.

i. Use Trypsin-EDTA (Cat No. CC-3232) to de-adhere the MSCs.

ii. Transfer the cells into a centrifuge tube. The size of the tube will depend on the total volume. Since the trypsin can remove surface proteins, you should incubate the cells for 1–2 hours to allow the proteins to be re-expressed; otherwise, you will get false negative data.

iii. Centrifuge the tube for 7–10 minutes at 300 *g*. The time will depend on the total volume. Larger volumes will require more time to pellet the cells.

iv. Aspirate the supernatant and resuspend the cell pellet at a concentration of $5 \times 10^6$ to $10^7$ cells/ml in BD Pharmingen™ Stain Buffer (Cat No. 554656).

v. Label tubes as shown below:

  a. Tube 1 – cells alone

  b. Tube 2 – 5 µl each of the Positive Isotype Control Antibodies (BD™ Mouse IgG1 PE, BD™ Mouse IgG1 PerCP-Cy™ 5.5, BD™ Mouse IgG1 APC, and 5 µl each of the Negative Isotype Control Antibodies (BD Pharmingen™ FITC Mouse IgM, κ Isotype Control, BD Pharmingen™ FITC Mouse IgG1, κ Isotype Control) (Note 4)

  c. Tube 3 – 5 µl of PerCP-Cy™5. Mouse Anti-human CD73

  d. Tube 4 – 5 µl of FITC Mouse Anti-human CD44

  e. Tube 5 – 5 µl of APC Mouse Anti-human CD105

  f. Tube 6 – 5µl of PE Mouse Anti-human CD90

  g. Tube 7 – 5µl each of PE Mouse Anti-human CD90, APC Mouse Anti-human CD105, PerCP-Cy™5 Mouse Anti-human CD73 (this is the positive cocktail). Add 5 µl each of the FITC Mouse Anti-human CD34, 45, 11b, 14, and FITC HLA-DR (this is the negative cocktail) (Note 6).[6]

vi. To tubes 1–7, add 100 µl of the prepared cell suspension from step iii. Incubate tubes in the dark on ice for 30 minutes.

vii. Centrifuge the tubes at 300 *g* for 7 minutes; aspirate the supernatant, and resuspend the cells in BD Pharmingen™ Stain Buffer (FBS). Repeat this step twice.

viii. Resuspend at 300–500 µl in BD Pharmingen™ Stain Buffer (FBS) or 1X Washing/Staining Solution (1 × PBS, 1% FCS, and 0.09% sodium azide) for the final suspension.

ix. Analyze cells on a flow cytometer. Use tubes 1–6 as controls to set up the cytometer (i.e., compensation):

a. Use the single-stain positive tubes (tubes 3–6) to set the compensation and acquisition parameters. Your flow cytometer's manufacturer will have manuals and methods on how to set the instrument's settings and properly compensate before running samples.

b. Tubes 1 and 2 can be used as controls for gating and negative standardization. Tube 1 is a population of unstained sample cells and will show the natural auto-fluorescence cells in the absence of antibodies. Tube 2 is an isotype control of both positive and negative cocktails.

c. Once the first six tubes have been used for standardization, tube 7 can be run to yield the target data of interest. Ideally, the three positive markers CD90-PE, CD105-APC, and CD73-PerCP-Cy™5 should have fluorescence above the threshold set during these previous standardization steps.

d. The five markers for which MSCs should not be positive (CD34, CD45, CD14, CD11b, and HLA-DR) are all targeted by FITC-conjugated antibodies. In the current procedural design, if the acquisition of tube 7 yields FITC fluorescence above the threshold set in the standardization steps, then there is expression of non-MSC markers and the sample population may not be a true or pure mesenchymal stem cell phenotype. You may repeat steps iv–viii only instead creating five tubes in step iv containing 5 µl of one of the FITC negative markers and run each tube separately in order to determine which exact negative marker is being expressed.

e. Quantitative analysis of the acquisition data from flow cytometry can be done on FloJo™ or similar flow cytometry analysis software, as per your cytometer manufacturer's recommendation.

## 15.3.2 Tri-lineage differentiation capacity of the isolated MSCs

The differentiation capacity of MSCs is assessed with confluent cells using the induction media for Adipogenic, Osteogenic, and Chondrogenic lineages. Similar to phenotyping, multilineage differentiation should begin after passage 3. The specific media are commercial kits, e.g., Lonza, StemCell Technologies, Thermofischer, or Sigma-Aldrich.

### 15.3.3 Neuronal transdifferentiation of the MSCs [13]

### 15.3.3.1 Neuronal induction of hMSC

i. Culture MSCs from BM aspirates using complete DMEM media.

ii. When cells reach approximately 70%–80% confluence, trypsinize and subculture MSCs in 60-mm Falcon 3002 Petri dishes or on round Fisherbrand microscope selected cover glass are placed in 35 mm Falcon 3001 dishes (Fischer Scientific, Springfield, NJ). Characterization of neuronally induced MSCs will require the cells to be placed on glass cover slips. Let the cells achieve 20%–40% confluence.

iii. Replace DMEM media with neuronal induction medium (NIM), which comprises Ham's DMEM/F12, 2% FBS (Sigma), B27 supplement, 20 mM RA, and 12.5 ng/ml bFGF.

iv. Allow the MSCs to culture in the new media for a maximum of 12 days. Do not change the media during the induction period.

v. Noticeable changes toward a more neuronal morphology begin to be visible after the sixth day of incubation in the neuronal media. Cells should ideally show neurite-like projections after 12 days but will not have axonal structures as pronounced as primary neurons.

### 15.3.3.2 Characterization of neuronal cells derived from hMSC Immunofluorescence

i. Establish uninduced (D0) and induced (up to D12) MSCs on glass coverslips placed in 35 mm culture dishes. Create enough coverslips to accommodate two pairs of fluorescence target designs (Tau/MAP2 and NeuN/GFAP) plus untreated controls and secondary antibody controls.

ii. Wash the cells with PBS (pH 7.4) and fix with 3.7% formaldehyde for 5 minutes.

iii. Permeabilize the cells in 1% Triton-X.

iv. Incubate overnight at 4 °C with the following antibodies: goat anti-Tau, rabbit anti-MAP2 at final concentrations of 1/100, and NeuN mAb and mouse anti-GFAP at final concentrations of 1/500. Dilute antibodies in 0.1% bovine serum albumin (BSA)/PBS.

v. Develop primary antibodies with secondary FITC-goat anti-mouse and goat anti-rabbit, and PE-goat anti-rabbit and rabbit anti-goat antibodies, at final concentrations of 1/500.

vi. Dilute secondary antibodies in 0.1% BSA/PBS and incubate for 2 hours in the dark at room temperature.

vii. Use cells labeled with PE- and FITC-non-immune IgG as isotype controls.

viii. Following labeling, counterstain cell nuclei and/or cytoskeletons with 300 nM DAPI and/or 6.6 M Texas Red phalloidin, diluted in 0.1% BSA/PBS, respectively.

ix. Transfer coverslips immediately to glass cover slides.

x. Examine on a three-color fluorescent microscope, e.g., Nikon Instruments Inc., Melvelle, NY.

## 15.4 Notes

Note 1: Trypsin is a proteolytic enzyme used to cleave polypeptides. It is used for short exposure of the cells to degrade the proteins that allow adherent cells attached to surfaces. It is only necessary for adherent cell lines and, when used, can destroy surface markers on the cell membrane. Thus, it is possible that your membrane protein of interest might not be expressed after trypsinization. To ensure that the protein gets re-expressed, incubate the cells at 1–2 hours before beginning to label the cells.

Note 2: The stain buffer can be formulated in-house instead of purchasing as pre-made. The solvent will be PBS, and then the FBS and sodium azide can be added to reach the specified percentages. Note that sodium azide is a cytotoxic compound used to preserve antibody–antigen binding and should not be used if you wish to preserve the viability of cells.

Note 3: A variety of conjugated antibodies both for positive visualization and control are needed for this protocol. You could get kits with the proper component of antibodies. These kits may be utilized and users may follow the manuals and protocols provided by the manufacturer of such kits. This chapter provides a list of antibodies that can be used in the absence of a kit.

Note 4: Positive antibodies are those used to identify MSCs by phenotype. As per the literature, MSCs should be positive for specific antibodies with the most commonly used antibodies such as anti-CD73, anti-CD90, and anti-CD105. If you do not have the option of three different fluorochromes to evaluate the three antibodies in the same tube, you may label each in separate tubes.

Note 5: There are several key markers that MSCs have not been reported in the literature; particularly, markers of hematopoietic cells. In general, CD45 will suffice.

Note 6: The negative isotype controls are conjugated with the same fluorochromes. This provides a standard against which a minimum threshold can be set for what is autofluorescence of non-specific antibody background binding.

Note 7: The positive isotype controls are conjugated with the same fluorochromes as the positive markers, and only their antigen targeting regions are designed to target to proteins not naturally expressed in humans/mammals. This provides a standard against which a threshold can be set for what is positive expression and what is autofluorescence of non-specific antibody background binding.

## References

[1] Ullah I, Subbarao RB and Rho GJ. Human mesenchymal stem cells - current trends and future prospective. Biosci Rep. 2015; 35(2): e00191.

[2] Caplan AI. Mesenchymal stem cells. J Orthopaedic Res. 1991; 9(5):641–650.

[3] Dominici M, Le Blanc K, Mueller I, Slaper-Cortenbach I, Marini F, Krause D, Deans R, Keating A, Prockop D and Horwitz E. Minimal criteria for defining multipotent mesenchymal stromal cells. The International Society for Cellular Therapy position statement. Cytotherapy. 2006; 8(4):315–317.

[4] T LR, Sánchez-Abarca LI, Muntión S, Preciado S, Puig N, López-Ruano G, Hernández-Hernández Á, Redondo A, Ortega R, Rodríguez C, Sánchez-Guijo F and del Cañizo C. MSC surface markers (CD44, CD73, and CD90) can identify human MSC-derived extracellular vesicles by conventional flow cytometry. Cell Commun Signal. 2016; 14:2.

[5] Anastasio A, Gergues M, Lebhar MS, Rameshwar P and Fernandez-Moure J. Isolation and characterization of mesenchymal stem cells in orthopaedics and the emergence of compact bone mesenchymal stem cells as a promising surgical adjunct. World J Stem Cells. 2020; 12(11):1341–1353.

[6] Robert AW, Marcon BH, Dallagiovanna B and Shigunov P. Adipogenesis, Osteogenesis, and Chondrogenesis of Human Mesenchymal Stem/Stromal Cells: A Comparative Transcriptome Approach. Front Cell Dev Biol. 2020; 8:561.

[7] Hernandez R, Jimenez-Luna C, Perales-Adan J, Perazzoli G, Melguizo C and Prados J. Differentiation of Human Mesenchymal Stem Cells towards Neuronal Lineage: Clinical Trials in Nervous System Disorders. Biomol Ther. 2020; 28(1):34–44.

[8] Kanaan NM and Grabinski T. Neuronal and Glial Distribution of Tau Protein in the Adult Rat and Monkey. Front Mol Neurosci. 2021; 14:607303.

[9] Gusel'nikova VV and Korzhevskiy DE. NeuN As a Neuronal Nuclear Antigen and Neuron Differentiation Marker. Acta Naturae. 2015; 7(2):42–47.

[10] Hendrickson ML, Rao AJ, Demerdash ON and Kalil RE. Expression of nestin by neural cells in the adult rat and human brain. PLoS One. 2011; 6(4):e18535.

[11] Potian JA, Aviv H, Ponzio NM, Harrison JS and Rameshwar P. Veto-Like Activity of Mesenchymal Stem Cells: Functional Discrimination Between Cellular Responses to Alloantigens and Recall Antigens1. J Immunol. 2003; 171(7):3426–3434.

[12] Kang HS, Habib M, Chan J, Abavana C, Potian JA, Ponzio NM and Rameshwar P. A paradoxical role for IFN-γ in the immune properties of mesenchymal stem cells during viral challenge. Exp Hematol. 2005; 33(7):796–803.

[13] Greco SJ, Zhou C, Ye JH and Rameshwar P. An interdisciplinary approach and characterization of neuronal cells transdifferentiated from human mesenchymal stem cells. Stem Cells Dev. 2007; 16(5):811–826.

# 16

## Cryopreservation of Stem Cells

**Lauren S. Sherman[1,2] and Pranela Rameshwar[1]**

[1]Department of Medicine, Rutgers New Jersey Medical School, USA
[2]Rutgers School of Graduate Studies at New Jersey Medical School, USA
**Corresponding Author:** Pranela Rameshwar, Rutgers New Jersey Medical School, MSB, USA
E-mail: rameshwa@njms.rutgers.edu
**Disclaimer:** The authors declare no conflict of interest.

## Abstract

Cryopreservation permits long-term storage of primary cells and cell lines for future use. The cryopreservation process must achieve >90% viability upon defrosting. Furthermore, it is important that the method is performed in a manner that does not cause the cells to burst during the process. This method provides instructions for freezing mesenchymal stem cells (MSCs) from various tissue sources, with notes for adjusting the method to other stem cell types. The described method for MSCs is similar for other stem cells except for mobilized peripheral blood stem cells, which is described in this volume.

## 16.1 Introduction

Cell cryopreservation is the process through which cells are frozen to a very low temperature in a controlled environment to minimize cell death. The most common temperature environment used is gas phase liquid nitrogen; this temperature is both easy to safely reach in a standard research laboratory (with applicable safety procedures) and to keep cells metabolically inactive for extended storage. When working with primary cells, it is preferable to freeze cells at an early passage number such that they can be adequately expanded upon defrosting for future use.

123

The primary hurdles that must be overcome in cryopreservation are ice crystal formation, damage to the cell membrane during the freezing and thawing processes, and osmotic shock [1]. Advances in cryoprotective agents and controlled rate freezing devices have enhanced the efficiency of long-term cell storage. Dimethyl sulfoxide (DMSO) is one such cryoprotective agent: this cell membrane-permeating sugar enters the cell, increasing viscosity and decreasing the freezing temperature within the cell [2]. Modern controlled rate freezing devices can be used to further enhance freezing efficiency and viability, decreasing the temperature at a controlled rate without exposing the cells to a sudden temperature shock. This method, however, focuses on mesenchymal stem cells (MSCs) and is applied for all stem cells, except mobilized peripheral blood stem cells (MPBSC), which is described in Chapter 18.

## 16.2 Materials

*All materials should be sterile and used under aseptic conditions unless stated otherwise. Media and cell numbers are listed for cryopreservation of human mesenchymal stem cells (MSCs).*

1. Dulbecco's Modified Eagle Medium, with high glucose and sodium bicarbonate, without L-glutamine and sodium pyruvate (DMEM; Sigma Aldrich D5671 or equivalent; Note 1)

2. Fetal bovine serum (FBS; Note 2)

3. Dimethyl sulfoxide (DMSO)

4. Cryovials

5. Control rate freezing container (Note 3; e.g., Mr. Frosty™ Freezing Container, ThermoFisher Scientific Cat. No. 5100)

6. Isopropyl alcohol

7. Access to a −80 °C freezer

8. Access to a liquid nitrogen storage tank containing liquid nitrogen

9. Gloves suitable for use with low temperatures

10. Safety glasses

## 16.3 Methods

1. Prepare freezing media. Two solutions will be used:

   A. Solution A: DMEM supplemented with 20% FBS (e.g., 8 ml DMEM and 2 ml FBS). You may use any other tissue culture media since the actual expansion media is not required during cryopreservation.

   B. Solution B: DMEM supplemented with 20% FBS and 20% DMSO (6 ml DMEM, 2 ml DMSO, and 2 ml FBS). See "A" for information on media.

2. Ensure that the controlled rate freezing container is prepared. In the case of the Mr. Frosty™ Freezing Container, the isopropyl alcohol should be at the level of the "fill line."

3. Collect non-adherent cells following trypsinization. The incubation time will depend on the cell type. Refer to the specific methods for the time documented during passaging.

4. Resuspend the cells in a small amount of Solution A. You want to use the least amount of this solution because if it is too diluted, you will have to centrifuge to get the appropriate concentration for cryopreservation (see Method 18 for cell count).

5. Resuspend the cells at $1 \times 10^6 - 5 \times 10^6$ cells per ml in Solution A.

6. Prepare cryovials, calculating the number of vials needed. This will be based on the total cell count and the desired number of cells/vial (e.g., 1 ml cryovial would be good for $5 \times 10^5 - 1 \times 10^6$ MSCs, while a 2 ml cryovial would be good for double the amount of cells). Each cryovial should be labeled with the cell type, passage number, date, and any relevant modifications and/or relevant identifiers. If your protocol, approved by the Institutional Review Board (IRB), states no identifier, you must adhere to the protocol when labeling the tubes.

7. Adjust the total volume of solution A so that the concentration of the cells is double that of what will be needed for the 1 ml final solution. The addition of solution B will dilute the cells so that the final concentration is readjusted (Note 4).

8. Dropwise add equal volume of Solution B to the resuspended cells. Gently shake the tubes while adding the solution. The resuspended

cells should now be resuspended in 50% each Solution A and Solution B.

9. Immediately transfer the resuspended cells to the labeled cryovials and place the cryovials into the controlled rate freezing apparatus.

10. Immediately transfer the freezing apparatus to the −80 °C freezer. After 12–24 hours, transfer the cells to liquid nitrogen for long-term storage (Note 5). Appropriate safety equipment including temperature-resistant gloves and safety goggles should be used when handling liquid nitrogen to prevent burns.

## 16.4 Notes

Note 1: If using a cell type requiring a base culture media other than DMEM, select the media accordingly. However, it is not necessary to use the same culture media since the cells are in the particular media only during storage. Although you may use any available media, ensure that you use the same media in both solutions A and B.

Note 2: If the cells are to be used in patients, a non-xenogeneic serum equivalent, such as human serum albumin, can be used in place of FBS. For clinical applications, cryopreservation has to be done in a laboratory that has been certified for good manufacturing procedure (GMP). Also, you should use a certified control freezing method.

Note 3: Most human cells, including stem cells, are best frozen at a controlled rate of −1 °C/minute – as accomplished in a Mr. Frosty™ Freezing Container. Other control rate freezing containers or machines can be used in its stead. This freezing rate may differ for other cell types.

Note 4: Solution B should be added slowly to ensure that the cells are not shocked by a sudden influx of high concentrations of DMSO. It is recommended to add Solution B dropwise while agitating the tube of resuspended cells.

Note 5: Once Solution B has been added, the cells should be frozen as quickly as possible to prevent DMSO-mediated toxicity. Avoid keeping the cells for >24 hours at −80 °C; otherwise, you will have reduced cell viability.

# References

[1] Jang TH, Park SC, Yang JH, Kim JY, Seok JH, Park US, Choi CW, Lee SR and Han J. Cryopreservation and its clinical applications. Integr Med Res. 2017; 6(1):12–18.

[2] Pegg DE. The history and principles of cryopreservation. Semin Reprod Med. 2002; 20(1):5–13.

# 17

# Cryopreservation of Mobilized Peripheral Blood Stem Cells

Seda Ayer[1], Khadidiatou Guiro[1], Steven S. Greco[1], Oleta A. Sandiford[1], Lauren S. Sherman[1,2], Garima Sinha[1,2], Nykia D. Walker[1,2], and Pranela Rameshwar[1]

[1]Department of Medicine, Rutgers New Jersey Medical School, USA
[2]Rutgers School of Graduate Studies at New Jersey Medical School, USA
**Grant:** The Bosarge Family Foundation supported this work.
**Corresponding Author:** Pranela Rameshwar, Department of Medicine, Rutgers New Jersey Medical School, USA.
E-mail: rameshwa@njms.rutgers.edu

## Abstract

Among the techniques described in this book is a chapter (#16) describing how to cryopreserve cell lines and primary cells. This chapter pertains to a relatively small number of cells, such as the amount expected during passaging of cell lines and from primary sources such as umbilical cord blood and bone marrow aspirate. However, our experience with a large number of cells from human mobilized peripheral blood (MPB) indicated that cryopreservation of these large number cells would require a modified method of cryopreservation. In general, MPB cells are indicated for hematopoietic stem cell transplantation. However, MPBs, which contain hematopoietic stem cells and progenitors, are excellent sources of hematopoietic cells for research purposes. This chapter describes the method to cryopreserve the relatively large number of hematopoietic cells expected from one donation of MPB. Donor MPBs may be obtained from any approved commercial site or discarded samples from a hematopoietic transplant facility. It is important that the investigator must obtain approval from the institutional review board prior to soliciting the samples.

## 17.1 Introduction

The methods to cryopreserve cells are mostly similar when the total number of cells is relatively small, i.e., less than $10^9$. We have reported on the use of mobilized peripheral blood (MPB) from young and aged individuals to investigate hematopoietic restoration in older individuals [1]. This required collecting MPBs from young and aged donors, which had to be cyropreserved for later use of the cells. This chapter provides a stepwise process that ensures effective storage of the MPBs.

MBP is the process by which hematopoietic stem and progenitor cells are mobilized by granulocyte colony stimulating factor (G-CSF) or other similar agents into the peripheral blood system [2]. This allows for a less invasive system when harvesting hematopoietic cells for allogeneic and autologous transplants. However, MPB using a similar procedure to harvest the cells may be used in experimental studies. In other cases, when a bone marrow (BM) recipient passes away before transplantation, the stored samples may be discarded. These samples may be used for research purposes, following the institutional agreed process. Scientists who are engaged in using MPB for research will need to agree if they should use an approved commercial source to acquire the samples or to use their institution if appropriate. Regardless of the method to acquire the MPB, you must have an institutional approval (Note 1).

## 17.2 Method

### 17.2.1 Reagents

#### 17.2.1.1 Cryopreservation media

Note: You will need ~200 ml of *freshly prepared* cryopreservation media per MPB sample to achieve a final volume of 400 ml. Chill the media on ice or at 4 °C.

- Human serum albumin (HSA), Irvine Scientific

- Dimethyl sulfoxide (DMSO), Fisher Scientific

- Sterile normal saline (USP grade)

Chill cryopreservation media in a 500 ml sterile bottle. The temperature of the bottle should be adjusted to 4 °C in a beaded ice bucket. The cryopreserved media comprise sterile normal saline, 3.6% has, and 20% DMSO (2× media).

- If you are not using the HSA in a liquid, reconstitute in saline and sterile filter through a 0.2 μ membrane.

### 17.2.1.2 Wash buffer (optional if you plan to wash the MPB cells)

- 1× Tissue culture grade PBS (pH 7.4) or sera-free tissue culture media (any type).

- 2% has.

### 17.2.1.3 Tissue culture disposables

- 2 or 5 ml cryopreservation tubes (any commercial source).

- 50 ml tissue culture polystyrene tubes (any commercial source).

### 17.2.1.4 Thawing of the cells
**Reagents:**

- DNase 1, RNase-free, 10,000 U, Sigma if thawing one or two vials. Use Worthington's DNase I, RNase-free lyophilized reagent for large-scale preparation.

- RPMI 1640 + 5% HSA + 30 U DNase I

  - If you are using DNase at 10,000 U total. This arrives with 10× buffers.

  - Dilute the DNase to ensure the buffer is 1×:

    - For example, 1000 µl buffer added to 9.8 ml sterile endotoxin-free $H_2O$ and 100 µl DNase $\Xi$ 1× buffer.

    - Aliquot in at 2 ml tube. This should be sufficient for five 100 mm plates, each total volume = 10 ml.

### 17.2.1.5 Equipment

- CryoMed Freezer, Thermoscientific.

### 17.2.2 Samples

a. You may request MPB from an approved site, e.g., HemaCare (see Note 1). It is likely that your lab will be at another state from the commercial source. You should try to select a source for the MPB as close as possible. This will prevent long delay, which could compromise cell viability and/or cell functions. It is likely that the samples will be shipped overnight at 4 °C. Try to avoid travel time that would exceed 16 hours post-collection.

b. The collection may arrive in two bags with each containing 200 ml.

### 17.2.3 Preparation of cells for cryopreservation (e.g., method to cryopreserve one bag containing 200 ml)

a. Label 200 cryotubes the evening before the sample arrives or early, prior to the sample arriving in the lab. If labeling the previous night, leave the tubes in the tissue culture hood with the blower in the "On" position. To be prudent, leave the UV light on overnight for added sterility on the outside of the tubes.

- Include covered boxes under the UV light. Open the boxes to sterilize the cover and box.

b. Add the tubes to the box and incubate in the refrigerator.

c. Spray the MPB bag with 70% alcohol; wipe off the alcohol with sterile gauze; cut the top of the bag with a sterile scissors.

d. Transfer 40 ml of the MPBs in five 50 ml tubes (see Note 2).

e. Pellet the sample at 300 $g$ at 4 °C for 10 minutes.

f. Remove 50% of the supernatant from each tube (~20 ml).

g. Using a 10 or 25 ml pipette, carefully remove the other 20 ml of supernatant. Transfer to a sterile 500 ml glass bottle placed in a bucket of cold ice beads (the bucket should be wiped with alcohol).

h. Tap the tube to resuspend the cells.

i. (Optional) Wash the cells once with cold wash buffer or proceed to cryopreservation (see Note 3).

j. Resuspend the cell pellets in ~20 ml of the saved media from the 500 ml bottle.

k. Transfer the cells into the 200 ml supernatant in the bottle.

l. Slowly add 200 ml of chilled cryopreservation media (1:1 ratio, 400 ml final volume). Gently shake the bottle while adding the cryopreservation media.

m. Quickly dispense the cells into the pre-labeled 2 ml cryopreserved cells. To efficiently accomplish this, it is better to have three persons assisting with this step – one person will open the vials, the other will add 2 ml/vial (100 × 10$^6$ cells/vial or 50 × 10$^6$/ml) who will pass the vials for closure by the third person (see Notes 4 and 5). The third individual will intermittently move the vials in a sterile rack at 4 °C.

n. Once all the vials are filled with the cell suspension, immediately move them to the controlled rate freezer at −1 °C/minute until the temperature reaches −100 °C.

o. Transfer the vials to liquid nitrogen.

### 17.2.4 Thawing cryopreserve cells

a. Thaw one vial of cryopreserve cells through rapid shaking of the tube in a 37 °C water bath. Once thawed, spray the vial with 70% ethanol and then proceed to the laminar flow hood.

b. Quickly add the contents of the vial dropwise into 10 times more volume of pre-warmed RPMI containing 5% HSA and 30 U/mL DNase I. As a rule of thumb, this will be 30 μl DNase I/10 ml media.

c. Equilibrate the cells at 37 °C for 1–2 hours.

d. Pellet samples at 300 *g* for 10 minutes at room temperature (RT).

e. Wash once in PBS containing 2% BSA.

f. Resuspend the cells in RPMI 1640 with 2% with 2% BSA. Proceed with your assay.

## 17.3 Notes

Note 1: Do not request MPBs unless you have the Institutional Review Board (IRB) approval.

Note 2: If you fill the tube with wash buffer to almost 50 ml, you might require additional centrifugation to pellet the cells. You could examine the supernatant for residual cells, which will be indicated by cloudiness. If so, continue to centrifuge for an additional 5 minutes. *You are advised to add the 200 ml samples from the bag into five or six tubes. This will avoid loss of cells in the supernatant.*

Note 3: You could replace PBS with sera-free tissue culture media in the wash buffer. This could be costly but is likely to preserve viability.

Note 4: You should retrieve $50 \times 10^6$ mononuclear cells/ml.

Note 5: After combining the cryopreservation media to the cells, this will result in 1.8% HSA, 10% DMSO, and $50 \times 10^6$ cells/ml.

# References

[1] Greco SJ, Ayer S, Guiro K, Sinha G, Donnelly RJ, El-Far MH, Sherman LS, Kenfack Y, Pamarthi SH, Gergues M, Sandiford OA, Schonning MJ, Etchegaray JP, et al. Restoration of aged hematopoietic cells by their young counterparts through instructive microvesicles release. Aging. 2021; 13(21):23981–24016.

[2] Dreger P, Haferlach T, Eckstein V, Jacobs S, Suttorp M, Löuffler H, And WM-R and Schmitz N. G-CSF-mobilized peripheral blood progenitor cells for allogeneic transplantation: safety, kinetics of mobilization, and composition of the graft. Br J Haematol. 1994; 87(3):609–613.

# 18

---

## Manual Cell Count

---

**Andrew Petryna[1,2], Lauren S. Sherman[1,2], and Pranela Rameshwar[1]**

[1]Division of Hematology/Oncology, Department of Medicine, Rutgers New Jersey Medical School, USA
[2]Rutgers School of Graduate Studies at New Jersey Medical School, USA
**Corresponding Author:** Pranela Rameshwar, Department of Medicine – Division of Hematology/Oncology, Rutgers New Jersey Medical School, USA
E-mail: rameshwa@njms.rutgers.edu
**Disclaimer:** The authors have nothing to declare.

## Abstract

A research laboratory that handles multiple types of cells – primary and cell lines – might not want to use an automated system that would require standardization for different types of cells based on size. Furthermore, some laboratories might not have the financial resource to purchase an automated counter. In this regard, manual count could be done with a hemacytometer that could be reused. This chapter outlines a step-by-step method to manually count cells. We also outline the principle of the formula used to calculate the final cell count. Since the cells are counted visually on the microscope, the procedure is used for any size of cells.

## 18.1 Introduction

This chapter describes how cells are counted manually with a hemacytometer. Laboratories might be conducting cell counts as a minor part of their studies and therefore should omit the expensive automated counter. Another advantage of the manual count is the versatility to work with cells of differing size. This is overcome with the microscopic visualization of the cells. Laboratory personnel are advised to practice using the

135

hemacytometer and to compare their counts with the outcome of a more experienced personnel.

## 18.2 Prerequisite Skills

- Comfortable with simple microscopy

## 18.3 Reagents

- Turk's solution (2% acetic acid):

  | Crystal violet (Note 1) | 10 mg |
  | Glacial acetic acid | 2 ml |
  | Distilled water | 98 ml |

- Hemacytometer and cover slip (ThermoFisher)

## 18.4 Method

Cell dilution:
Cells may be counted as a diluted or undiluted solution. This will depend on the concentration of cells/ml. In cases where you have small number of cells without red blood cells (RBCs), you would not want to lose cells. In this case, count undiluted.

 If RBCs are present, eliminate them to get an accurate count of the white blood cells (WBCs). Dilute at 1/10 in Turk's. This will lyse the RBCs without affecting the white cells.

Cell count:
- Add the coverslip on top of the hemacytometer. Fill the grid with the cells using a pipette tip or a capillary tube containing the cells. Simply place the tip in the grove of the grid and the fluid will fill the space.

- If you are counting white blood cells or cell lines, count cells in the four outer large squares. Use the middle square for red blood cells and platelets (Figure 18.1).

- If you choose to count cells resting on the top and right lines, be consistent and do the same in the other three squares of the hemacytometer.

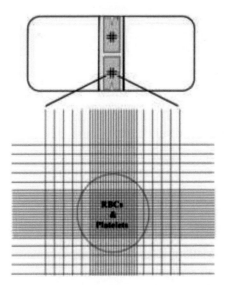

**Figure 18.1** Outline of a Neubauer hemacytometer. Nucleated cells are counted in the four corner squares. The cells on top of the edges are counted at the right and top sides in all four squares or at the left and bottom sides of the four squares.

Rule of thumb when diluting the cells:
 a. Total volume of cell suspension: ≤*1 ml*:

  • 90 µl diluent + 10 µl cell suspension (final 1/10).

 b. Total volume of cell suspension: ≥*1 ml*:

  • 0.45 ml diluent + 50 µl cell suspension (final 1/10).

Accuracy/rule of thumb:
The hemacytometer has four squares at the ends and one in the middle.

 i. Accuracy of the cell count is attained if the distribution of cells/square in the hemacytometer is about 50–100 cells.

 ii. To reiterate, be consistent when counting – if cells are counted on the top and right borders of the squares, you will count the same cells twice.

Errors:
Errors can be introduced in a number of ways. Common source of errors are:
 • During dilution:

  ○ Cell loss during pipetting.

     ○ Uneven cell suspension.

● Overfilling or underfilling the hemacytometer chamber.

● Counting of cells while the cells are in motion within the fluid in the hemacytometer.

● Random distribution of cells in the chambers. This can be compensated by counting several chambers and then using the average count/chamber.

<u>Cell enumeration:</u>

A. Quick estimation: (see "B" for the derivation of this formula. Make an effort to look at "B," at least once.

The number of cells counted in one large square is equivalent to the amount present in a volume of $10^{-4}$ ml.

Therefore, total number of cells/ml:

Number of cell in 1 sq. × $10^4$ × dilution factor.

B. Origin of the formula used in "A":

| | | |
|---|---|---|
| Side of square | = | 1.0 mm |
| Area | = | 1.0 mm² |
| Depth | = | 0.1 mm |
| Volume | = | 0.1 mm³ |

Let "**X**" be the number of cells counted in 1 sq.

Number of cells counted in 1 sq.  = "**X**"/0.1 mm³
(diluted suspension)            = "**X**" × $10^1$/1 mm³
                            = "X" × $10^1$ × $10^3$/cc (ml)
                            = "Y"

Total number of cells/ml
(undiluted suspension)       = "Y" × dil. factor (usually 1/10)
                      OR
     Number of cells in 1 sq. × $10^4$ × dil. factor

**Described in another manner:**

Number of cells counted $\times 10^4 \times$ dilution factor

Number of grids counted

$10^4$: conversion of cells/0.1 mm$^3$ (volume of grid = 1 mm $\times$ 1 mm $\times$ 0.1 mm (height))

## 18.5 Notes

Note 1: Crystal violet is optional. This is useful for those who have problems identifying the cells during cell count.

# Index

# About the Editor

**Pranela Rameshwar** is a tenured professor of medicine, Division of Hematology and Oncology at Rutgers New Jersey Medical School, Rutgers. She received her B.Sc. degree in medical microbiology from the University of Wisconsin at Madison and Ph.D. in biology from Rutgers University, New Jersey. She performed postdoctoral studies in hematopoiesis at New Jersey Medical School. Dr. Rameshwar's research interest in stem cell biology began with her doctoral thesis and continues with research on breast cancer dormancy, neural regulation of hematopoiesis, and clinical application of adult human mesenchymal stem cells. Her research continues to be funded by federal, state, and other agencies. Dr. Rameshwar has edited books and authored >200 publications, which include original articles, reviews, editorials and book chapters. Dr. Rameshwar initiated a certificate program in stem cell biology and directs four graduate level courses in stem cell biology.